PHILOSOPHY AND MEDICINE

Editors:

H. TRISTRAM ENGELHARDT, JR.

The Center for Ethics, Medicine, and Public Issues,
Baylor College of Medicine, Houston, Texas, U.S.A.

STUART F. SPICKER

School of Medicine, University of Connecticut Health Center,
Farmington, Connecticut, U.S.A.

VOLUME 21

THE PRICE OF HEALTH

Edited by

GEORGE J. AGICH

Southern Illinois University School of Medicine,
Springfield, Illinois, U.S.A.

and

CHARLES E. BEGLEY

School of Public Health, University of Texas Health Science Center,
Houston, Texas, U.S.A.

D. REIDEL PUBLISHING COMPANY

A MEMBER OF THE KLUWER  ACADEMIC PUBLISHERS GROUP

DORDRECHT / BOSTON / LANCASTER / TOKYO

Library of Congress Cataloging-in-Publication Data

The Price of health.

(Philosophy and medicine ; v. 21)

Based on a conference entitled "The price of health: economics and ethics in medicine," held at the School of Public Health, University of Texas Health Science Center, Houston, Tex., June 24–26, 1985.

Includes bibliographies and index.

1. Medical economics—Moral and ethical aspects—Congresses. 2. Medical care—Cost effectiveness—Congresses. 3. Medical ethics—Congresses. I. Agich, George J., 1947– . II. Begley, Charles E., 1947– . III. Series. [DNLM: 1. Economics, Medical—United States—congresses. 2. Ethics, Medical—congresses. W3 PH609 v.21 / W 74 P9455 1985]

RA410.A2P75 1987 174′.26 86–21877

ISBN 90–277–2285–4

Published by D. Reidel Publishing Company,
P.O. Box 17, 3300 AA Dordrecht, Holland.

Sold and distributed in the U.S.A. and Canada
by Kluwer Academic Publishers,
101 Philip Drive, Norwell, MA 02061, U.S.A.

In all other countries, sold and distributed
by Kluwer Academic Publishers Group,
P.O. Box 322, 3300 AH Dordrecht, Holland.

Printed in The Netherlands.

TABLE OF CONTENTS

PART IV / CONTROLLING COSTS/MAXIMIZING PROFIT: THE ROLE OF PROVIDERS

FOREWORD

Medicine, morals and money have, for centuries, lived in uneasy cohabitation. Dwelling in the social institution of care of the sick, each needs the other, yet each is embarrassed to admit the other's presence. Morality, in particular, suffers embarrassment, for it is often required to explain how money and medicine are not inimical. Throughout the history of Western medicine, morality's explanations have been consistently ambiguous. Plato held that the physician must cultivate the art of getting paid as well as the art of healing, for even if the goal of medicine is healing and not making money, the self-interest of the craftsman is satisfied thereby [4]. Centuries later, a medieval medical moralist, Henri de Mondeville, said: "The chief object of the patient . . . is to get cured . . . the object of the surgeon, on the other hand, is to obtain his money . . . ([5], p. 16).

This incompatibility, while general, is not universal. Throughout history, medical practitioners have resolved the problem – either in conscience or to their satisfaction. Some physicians have been so reluctant to make a profit from the ills of those whom they treated that they preferred to live in poverty. Samuel Johnson described his friend, Dr. Robert Levet, a Practiser of Physic:

> No summons mock'd by chill delay,
> No petty gain disdain'd by pride;
> The modest wants of ev'ry day
> The toil of ev'ry day supplied [3].

Many physicians have followed the same path. However, many others have felt no compunction about making a good living. The perennial satires of the physician attest to this. Chaucer's physician was

> Rather close in his expenses
> And kept the gold he won in pestilences.
> Gold stimulates the heart, or so we're told;
> He therefore had a special love of gold [1].

G. J. Agich and C. E. Begley (eds.), The Price of Health, vii–xi.

In Act III of Molière's 'Physician in Spite of Himself,' Sganarelle ruminates, "Medicine is the best trade of all, for whether you perform well or badly, you are always paid just the same." The advice of the medieval moralist quoted above can be taken either as cynicism or prudence: "The surgeon may assist the poor for the love of God, but the wealthy should be made to pay well" ([5], p. 15). The incompatibility between healing and making money is not exclusive to Western medicine. The seventeenth century Chinese physician-sage, Kunh T'hing-Hsein, said of his colleagues, "when they visit the rich, they are conscientious; when they deal with the poor, they are careless: this is the eternal peculiarity of those who practice medicine for financial gain and not out of humane care" ([6], p. 74).

However, the Western tradition has a history that reveals how the incompatibility gradually grew. In its most ancient origins the Hippocratic ethic placed no strictures against profiting from one's healing skills. Indeed, medicine was a trade and its practitioners were expected to earn their living by vending their skills and advice. The rare comment about providing free care for those in need can be interpreted simply as good business advice.

Above the ancient level is built the edifice of the Judeo-Christian ethic. Healing is a skill granted by God and should be dispensed as God's free gift to those in need. The Gospel image of the Good Samaritan appears in medical writing: as this compassionate person bound the wounds of a stranger and gave money for his care, so should the physician attend the sick out of compassion, and not for profit. In the late middle ages, the universities incorporated the teaching of medicine into their curriculum. A discipline was created that could be explained in written books and lectures and could be tested in examinations. The "healer" became the "doctor," and powerful professional bodies began to attain monopolies in the name of competence. Finally, in the 19th century, as measures for public health became feasible and enforceable, and as rulers, such as Germany's Bismarck, realized the importance, economically and politically, of a healthy workforce, state subvention of medical care appeared. Health insurance became part of the bargaining between labor and management. As care was paid for by someone other than the patient, the belief slowly emerged that persons had a *right* to health care.

Throughout this history, the Greek premise that medicine was a trade to be employed for profit fitted uneasily with the Judeo-Christian belief that medicine was a gift to be dispensed to the needy. The medieval creation of a learned profession and its monopoly did not mesh easily with the assertion of public right to the services of that learned profession.

As in many social institutions, these incongruities were occasionally noticed, but still tolerated. Practices to alleviate the tensions were devised. Wealthy physicians donated time to the poor; states and municipalities made arrangements of various sorts to provide care to the indigent; commercial insurance assumed, for a price, the economic burdens that ill health laid on individuals. Again and again, the preservation of professional monopoly was justified by the invocation of the ethics of patient and social welfare. These convenient adjustments relaxed the tension, provided needed services, and disguised the failures. Only in recent years has the adjustment begun to come apart. No longer does it seem possible to maintain the uneasy cohabitation of the old traditions by pallid moralisms. Medicine and health care are now obviously financial enterprises. The ethics of medicine is now clearly pressured by various economic constraints. Physicians, economists and philosophers must learn to understand each other better and replace their uneasy cohabitation with something approaching contractual clarity.

Economists have been studying medicine's use of money for the past few decades. Similarly, philosophers have been studying medicine's moral structure for about the same time. However, the economists have attended primarily to the modern manifestations of the Hippocratic and medieval traditions: they have studied how the doctor makes and uses money, and how the professional monopoly affects the flow of money. They have attempted to elucidate, as is the custom of their discipline, the economic incentives and disincentives in the social institution of health care.

The philosophers, on the other hand, have dwelt on the modern manifestation of the Judeo-Christian tradition and the 19th-20th century doctrine of rights. They have, in their fashion, studied the compassionate duty to care for the sick, even at personal cost. They have analyzed the concept of rights as it applied to health care — both the rights of patients in care and their right to receive care. So, for the few decades that economists and ethicists have studied the institution of health and medical care, they have considered different segments of its tradition.

Now they must focus on the entire tradition in its present complexity, seeing that self-interest, compassion, competence and justice are strands in a tightly woven rope that binds medicine, morality, and money to each other. Those who practice medicine, teach it, and advance its science must also understand the complexity of their tradition. Unfortunately, they seem to know their tradition less well than its outside observers. They have, in recent years, shown some interest in ethics and economics, but only a few have undertaken to address these subjects in depth. It is important that a tri-cornered conversation take place, as in this volume, in which each conversant understands the language and the methods of the other; the language of the ethical traditions of medicine and the language of the economics of health care must be melded.

The recognition that medicine, morality and money necessarily dwell together makes a book of this sort necessary. It represents the first generation of a discussion between the disciplines. It poses some of the issues familiar to economists and allows response in terms familiar to ethicists. Thus, cost-benefit analysis, long a part of the economist's methodology, is exposed to the philosopher's analysis and interpretation in light of the concept of justice. This sort of dialogue, only beginning in these pages and elsewhere, must be fostered. Failure to do so may harm medicine and health care ethics immeasurably. Medicine's traditional ethic, demanding attention to the needs of the sick, even the poorest of them, can be compromised by powerful economic forces. Ethics is a weak reed before the tempest of self-interest. It is even more enfeebled if the language and concepts of ethics are silently and subtly invaded by those of economics — indeed, the language and concepts of utilitarian ethical theory are congenial enough to modern economics. Even more serious, the values of the health care tradition can be slowly transmuted into the values of finance. Similarly, the language of ethics must not be used, as it has in the past, to disguise the monopolistic self-interest of the profession, but rather to support the values of the profession as a service, sensitive to the welfare and rights of persons.

It may be wise to remember, as one reads these pages, the advice of the Chinese sage about those "who practice medicine for money and not out of humane care". It is also wise to note the words of the distinguished economist, Victor Fuchs: "There are a number of questions that society must ask itself concerning the financing of 'caring'. How much should this service cost? Who should pay for it?" ([2], p.

66). The words of the sage and the economist, in juxtaposition, form the theme basic to this volume and to the conversation about medical morality and medical economics: that care does, indeed, have costs, but that costs must not destroy the humaneness of care.

April, 1986 ALBERT R. JONSEN

BIBLIOGRAPHY

[1] Chaucer, G.: *The Canterbury Tales*, Prologue.
[2] Fuchs, V.: 1974, *Who Shall Live?*, Basic Books, Inc., Publishers, New York.
[3] Johnson, S.: 'On the Death of Mr. Robert Levet, A Practiser of Physic'.
[4] Plato: *Republic*, §342.
[5] Reiser, S. J. *et al.* (eds.): 1977, *Ethics in Medicine: Historical Perspectives and Contemporary Concerns*, MIT Press, Cambridge, Mass.
[6] Unshuld, P.: 1979, *Medical Ethics in Imperial China*, University of California Press, Berkeley, California.

ACKNOWLEDGMENTS

This volume is mainly composed of papers prepared for a conference entitled 'The Price of Health: Economics and Ethics in Medicine' held at the School of Public Health, University of Texas Health Science Center, Houston, Texas, June 24–26, 1985. The conference was generously supported by the Center for Health and Manpower Policy Studies; a special note of thanks is due to Frank Moore, Ph.D., Director of the Center, for the hospitality extended to conferees.

The Department of Medical Humanities of Southern Illinois University School of Medicine and the Health Services Organization Program at the University of Texas School of Public Health additionally provided secretarial and editorial assistance. In particular, we wish to thank Lynne Cleverdon, Marilyn Flanigan, Linda Keldermans, Clara McAfee, Janie McBane, and Dortha Demas for their help.

INTRODUCTION

In recent years health policy discussion in the United States has been dominated by one major issue — the burgeoning cost of health care services. This issue has come to the forefront as people are increasingly priced out of the market, and government and business budgets are strained to subsidize access to such an expensive system. Since 1950, real per capita health care expenditures have nearly trebled ([1], p. 3). At present, national health care expenditures are nearly eleven percent of the Gross National Product (GNP), up from 4.4 percent in 1950 [6]. Hospital expenditures under the federal Medicare program for the elderly rose at the rate of 18 percent per year from 1977 to 1982 and states' expenditures for Medicaid assistance for the indigent have grown over 22 percent between 1981 and 1983, nearly twice as fast as state tax revenues [6].

The combination of rapidly rising costs, equally rapid advances in medical technology, and an aging population have raised the specter that medical care will soon have to be rationed, a seemingly anomalous prospect in the world's richest nation. In this regard, William Schwartz has observed in *Newsweek* magazine: "Either we will accept the continued rise of hospital costs that result from full exploitation of technological advances, or we will start to ration hospital care. And if it's the latter, we will then have to say to some people, 'Yes, a new liver would be good *for you*, but as a society we can't afford it'" [11].

It may surprise some to learn that medical care in the United States is already, and has always been, rationed. For example, John P. Bunker has pointed out that the poor and unemployed who cannot afford to pay for medical care and who do not qualify for public insurance often receive a lower quality of treatment, if they receive any at all ([3], p. 7). Although public hospitals and clinics are supposed to provide services to these people, many of those facilities are not funded well enough to deliver the same standard of care that exists in the private sector. Another form of rationing that occurs concerns the allocation of scarce resources such as intensive care beds, dialysis machines, or donated organs among the large number of patients who would benefit from them.

G. J. Agich and C. E. Begley (eds.), The Price of Health, xv–xxx.

What is new about the current prospect for rationing is that it involves limiting the consumption of "normal" health care resources by the middle class. A society that increasingly since World War II has come to expect the highest quality medical care, without regard to cost as a matter of entitlement or right, must now face the possibility of restrictions on the use of such ordinary medical resources as diagnostic tests, drugs, hospitals, and surgery. It is little wonder that policies designed to encourage such rationing have been met with considerable resistance.

Two general approaches to the so-called "health care cost crisis" have dominated policy discussions. The first approach is exemplified by the regulatory policies of the 1970s, which attempted to contain health care costs by imposing limitations on the availability of resources and by establishing guidelines for resource use, while at the same time striving to broaden access to health care services. The second approach has increasingly become popular in recent years; it relies on economic incentives or sanctions for cost-conscious clinical decision making and resource utilization. It is exemplified in programs such as the use of Diagnostic-Related Groups (DRGs) to pay hospitals on a prospective basis and procompetition based reforms such as Alain Enthoven's Consumer Choice Health Plan (CCHP) [5]. Both of these approaches involve explicit rationing of medical care as health care providers and patients are encouraged to reduce costs.

At the same time, discussion of health care in the United States has also focused on distributional inequities between rich and poor and between geographic areas [2]. Underlying this discussion is the belief that health care is a special social good which should be distributed solely on the basis of need. Corollary to this view is the belief that there is a special ethical obligation to provide high quality medical care even when the benefits are marginal or uncertain irrespective of the cost of care.

The seemingly incommensurable values reflected in these two viewpoints, as well as their philosophical roots and foundations, provide a context for the present volume. As the cost-containment effort intensifies, its conflict or agreement with other competing concerns or objectives must be made explicit. To do so requires analysis of the normative values involved in applying economic criteria to medical decisions, and how these criteria compare to other standards or norms which guide medical decision making. The papers in this volume attempt to provide such analysis by examining various questions concerned with the ethical and philosophical implications of economic analysis of health care: Is the provision of health care without regard to the efficient use of resources morally acceptable? Is it morally required? Is it morally wrong to regard health and health care as having economic value like other commodities or services? If health care is to be rationed, what role should economic criteria and ethical principles play in health care decision making? And finally, given the limitations on health care resources, what is the government's responsibility in securing access to care?

The burgeoning cost of health care is the corollary of a health policy that has evolved in the United States more as the result of broad historical and social developments and piecemeal political programs than of a comprehensive planning effort [13]. For that reason, it would be a mark of hubris to think that economic or philosophical analysis could affect directly the underlying cultural attitudes and political forces. Indeed, to even attempt to do so would presuppose that answers to difficult economic and philosophical questions were readily available. The essays collected in the present volume seek answers to some of the basic questions which any acceptable health policy will have to address. In this way it is hoped that the analyses and arguments offered will contribute to an understanding of some of the underlying issues that shape the present national dilemma.

The first section of this volume, entitled 'Medical Economics and Ethics: Some Theoretical Considerations', is comprised of two papers which critically analyze some of the main values that underlie theories of resource allocation in health care. In 'Economics and Allocation of Resources to Improve Health', Marc J. Roberts reviews the development of the three perspectives from which economists have considered the problem of allocating scarce resources to health care — namely, the human capital approach, willingness to pay, and cost-effectiveness analysis. He shows that although these perspectives have been developed within the same disciplinary tradition, they have quite different philosophical roots, and, more importantly for his purposes, they imply different views of the relationship between individuals and the state as well as different views regarding the objectives to which the state should be directed. Roberts is less concerned with discussing points of doctrinal dispute among economists than to pose the question: "To what extent are the various economic theories consistent with the ways in which health care resources have actually been allocated in the United States in recent years?" Robert's answer is instructive: there is a sharp divergence between economic theories of allocation of resources for health care and the practice by which health policies have been set.

Roberts attributes this divergence between theory and practice to limitations in the economic perspective which affect its ability to set adequate health policy directly. Economic arguments focus solely on consequences; accordingly, they leave out consideration of decision-making that ignores consequences such as actions that are done because they meet the needs of justice or charity irrespective of the outcomes they might produce. Nonetheless, Roberts contends, the economic point of view should not be set aside, because in a world of limited resources it is both imprudent and indefensible to ignore consequences altogether when allocating scarce resources. Economics thus reminds us that consideration of scarcity must figure in socially responsible decision making. The approach which Roberts proposes is not theoretical, but rather involves an appeal to the political process. He argues for a public discussion that would eschew rights and entitlement arguments, but instead would focus on the kinds of support we want to provide each other by virtue of our common citizenship.

In 'Economic Cost and Moral Value', George J. Agich suggests a view of the compatibility of economic and ethical concerns by critically discussing three alleged deontological principles of ethics which are often said to prohibit consideration of costs and benefits in medical decision making. Agich argues that the so-called traditional view of medical ethics, a view which holds that the physician is obligated by a unique principle of beneficence to provide the utmost benefit to a particular patient without regard for cost, is untenable. The traditional view of medical ethics focuses physician obligation on the welfare of individual patients and in general rejects obligations to society or other individuals outside the physician-patient relationship. In criticizing this view, Agich argues against two of its supporting propositions: first, health is an absolute moral good for physicians and, second, physicians are morally required to employ all the resources necessary to pursue health in the care of individual patients. In effect, Agich argues that the traditional view of medical ethics is more a description or explication of a practice style than a normative position built on defensible propositions.

The second view that Agich considers is the sanctity of life or respect for life view. This view holds that denying or withdrawing care on the basis of cost debases the fundamental sanctity of life and violates the principle which sees life as a moral object worthy of respect. Agich rejects this view on two grounds, namely, it fails to establish an adequate justification of its normative stance and the very effort to give substantive content to the respect for life principle in terms of a "right to life" requires consideration of the cost of medical care.

Finally, Agich considers the Kantian view that human life, particularly human rational life, is priceless. The Kantian view is that human persons have a dignity, but never a price or exchange value. However, while human persons cannot have exchange value insofar as they are the autonomous subjects of morality, human life as the bearer of empirical attributes such as skills, talents, or health is morally capable of being priced or exchanged insofar as the exchange does not violate autonomy. The view, which Agich defends, is derivative of Kant. Agich argues that health is an optional end or value; it is an instrumental good. He acknowledges the priceless character of human personal life as a basic axiom of moral theory, but argues that this does not preclude the pricing or differential valuing of various attributes or qualities of human life, one of which is health. Health, together with those things necessary for its attainment or preservation, are instrumental goods, goods worth pursuing in order to attain human welfare or happiness. The place of health in morality is central only insofar as health is freely pursued by persons.

Both Agich and Roberts suggest that medical ethics and economics, as traditionally understood and practiced, are themselves components of the controversies contributing to current health policy dilemmas. Agich shows that commonplace objections to the consideration of costs in medical care are without moral foundations. Roberts suggests, by reflecting on the historical development of economic modes of thinking, that there may be valid reasons for favoring health policies that conflict with economic appraisals.

The relationship between methods of economics, particularly cost-benefit and cost-effectiveness analysis, and ethics in health policy and medical decision making is addressed in the second section, entitled 'Costs and Benefits in Medicine: Some Philosophical Views'. Papers comprising this section are concerned with the philosophical assumptions underlying normative economic analysis. In 'Computing The Quality of Life', E. Haavi Morreim focuses on ethical questions associated with quality of life calculations in economic studies. She contends that the quality of life concept includes both subjective and objective senses which, because they are often inadequately distinguished, obfuscate much of the economic analysis of health care policies. Morreim distinguishes between the *objective characteristics* or facts of one's life, which include both the material circumstances which can be intersubjectively agreed upon and those judgments regarding the non-physical circumstances which are so widely shared as to constitute a kind of "social fact", and *subjective normative judgments* that arise from the valuing of individual subjects.

Morreim argues that the choice of an objective or subjective standard of quality of life should depend on the purpose of the program or policy, whether it aims to benefit society at large or to help individuals for their own sakes, or both. She shows how empirical measures of quality of life often fail to recognize the distinct, albeit related, notions of subjective and objective quality of life assessment. In those studies which recognize a difference, the difficulty lies in attempts to obtain valid measures. Morreim notes how subjective quality of life judgments can be problematic in clinical settings, especially in cases involving incompetents, and argues that, when available, subjective quality of life assessments nonetheless should assume prima facie moral importance in clinical settings if individual autonomy — even in its most restricted forms — is to be respected. Overall, Morreim's effort is to show how commonly used empirical measures of quality of life belie the complexity of quality of life assessment.

In 'CBA, Utilitarianism, and Reliance upon Intuitions', Peter S. Wenz instructively locates cost-benefit analysis in terms of utilitarian moral theory, specifically preference maximization utilitarianism. He indicates the theoretical strengths as well as the theoretical difficulties which CBA draws from utilitarianism. Wenz argues that concerns of distributive justice often conflict with actions justified in cost-benefit terms. Although use of the Pareto criterion avoids this consequence by requiring that policies be adopted only when there are no losers, it does so by making economic analysis irrelevant to most practical policy decisions [9]. The Kaldor-Hicks principle, which requires that losers be able to compensate winners, but not that they do so in fact, sanctions distributional inequity ([7], [8]). Wenz shows that modifications to cost-benefit analysis, such as weighting schemes designed to incorporate such distributional concerns, ultimately fail because the modifications can only be justified and employed by means of intuitions.

Wenz also argues that CBA fails to deal adequately with irrational desires. As a result, CBA can support conclusions that conflict strikingly with morality. Wenz, like Roberts, sees alternatives to the calculative rationality which CBA represents. Whereas Roberts looks to the political process, Wenz refers to a particular institution as an illustration, namely, the United States Supreme Court, which utilizes intuitive judgment guided by principles of interpretation based on the United States Constitution and legal precedent.

The criticism of cost-effectiveness and cost-benefit analysis offered by Morreim and Wenz is balanced by an argument by Paul Menzel in 'Prior Consent and Valuing Life' who sees the willingness to pay approach as a justified way to incorporate considerations of scarcity in medical care. Menzel argues that willingness to pay, which he treats as prior consent to risk of death or ill health, ought to be used in rationing health care services for three main reasons. First, the willingness to pay to reduce risk to life indicates a valuation of life and not just the freedom from risk; second, willingness to pay money is a good indicator of a real value of life, even though such value is only monetary value; and third, there is no serious problem in making different monetary valuations of the same life under different circumstances, such as in low risk situations versus high risk situations.

Menzel develops his argument by reference to historical examples in which willingness to pay has been utilized in contexts other than health care. He notes, for example, that life insurance has been accepted as allowing one to benefit from death without corrupting respect for life. Similarly, Menzel examines the practice of tort law which often establishes a monetary valuation of human life to compensate for wrongful death. Menzel argues that such utilization of the willingness to pay approach does not involve special moral problems for society. Of course, this is not so much a theoretical justification of willingness to pay as it is an argument based on observation of cultural norms.

Menzel argues that society is justified in providing an individual with that amount of care which the individual has consented to at some prior time either through explicit insurance arrangements or inferences drawn regarding the individual's aversion to risk in other contexts. Menzel acknowledges that actual prior consent is different from presumptive prior consent, which is extrapolated from data concerning what people are willing to risk in order to save money, but he does not see significant moral problems associated with the use of presumptive prior consent in evaluating health care. Menzel also discusses in detail the kind of prior consent that is morally required, as well as problems related to age and disease frequency. Although he relies on empirical evidence or inferences drawn from empirical evidence throughout his analysis, Menzel also cautions regarding the uncritical use of empirical data. Nonetheless, he argues that the concept of prior consent provides an ethically acceptable basis for making difficult allocational and distributional decisions in health care.

The common thread linking the papers on cost-benefit and cost-effectiveness analysis is the focus of Robert Audi's paper, entitled 'Cost-Benefit Analysis, Monetary Value, and Medical Decision'. Audi examines the decision-theoretic tradition which underlies these modes of analysis and discusses in detail several theoretical and ethical problems associated with that model. Audi finds fault with the model in the following respects. First, he argues that there are many instances in which one cannot justifiably form the appropriate range of probability beliefs that is required by the model. Second, he contends that the model makes no provision for identifying or discounting irrational and false beliefs or irrational evaluations (a major concern of the Wenz paper). Third, while the model does seem adequate to provide necessary and sufficient conditions for deciding *what choice is rational* for an individual, it does not provide such conditions for an individual *to rationally choose* or decide on a course of action, because there may be other conditions of choice besides those specified by the decision-theoretic model. This latter distinction figures importantly in Audi's analysis and provides the basis for his argument that ethical considerations bear significantly on an individual's choice of a course of action beyond the fact of whether the individual's decision conforms to the norms of rationality required by the decision-theoretic model.

Audi considers in detail some problems associated with the concept of tacit consent (or, in Menzel's terms, prior consent). By focusing on cases in which informed consent can be regarded as providing a moral warrant for action, Audi argues that more than rationality is required. In this regard, Audi draws an illustrative analogy with belief and valuation. To speak of how much one values one's life is to speak at best imprecisely. In order to specify a person's valuations, it is important to use terminology discriminating enough to reveal the content of an individual's propositional attitude. Talk of valuing one's life thus tends to cloak a variety of specific valuations connected with one's sense of what is worthwhile in life.

Audi argues that to seriously regard the role of an individual's monetary valuations from a moral point of view implies, at the least, that health policy discussions be clear regarding the distributional issues at stake. Regarding the question of whether there is a role for people's apparent pricing of their lives in determining health policy, Audi argues affirmatively. However, Audi thinks that public debate and voting processes would provide a better grounding.

Papers in the third section, entitled 'Economics and Ethics in Health Policy', consider directly two important health policy issues: the application of technology to diagnose, treat, and prevent genetic disease and some problems associated with defining and achieving a so-called basic or decent minimum standard of care. In 'Intervention Against Genetic Disease: Economic and Ethical Considerations', J. Michael Swint and Michael M. Kaback analyze one area of growing importance for the development of cost-effective health care policy. Rather than providing a theoretical analysis or justification, these authors primarily are concerned to clarify the complexities associated with a health policy problem that in large part is shaped by technological developments. Their analysis aptly illustrates how important empirical evidence is to any rational and informed public policy making and how technological innovations alter not only the cost-benefit considerations associated with particular genetic diseases, but the ethical considerations as well.

Swint and Kaback argue that ethical concerns such as protecting autonomous decision making, preventing harm, and maintaining confidentiality of test results alternatively agree or conflict with policy indications given by economic criteria. Hence, it is important that policy decisions regarding genetic screening and counseling be determined with full awareness of the economic, ethical, and health consequences.

Swint and Kaback focus on three illustrative diseases: Tay-Sachs disease, Down's syndrome, and Neural Tube Defects. These diseases present increasingly difficult conflicts for economics and ethics. The authors address the conflicts by reflecting on current social attitudes regarding, for example, the ethical acceptability of interventions designed to prevent or treat these diseases. Their discussion of the clinical presentations, available diagnostic and treatment options, and individual and social costs associated with alternative approaches illustrates the impracticality of a uniform policy with respect to genetic intervention. Differences in, for example, severity of the disease, the at-risk population, and availability of alternative modes of treatment comprise an essential knowledge-base which informed participants in policy debates will have to consider.

In 'Ethical Reflections in Genetic Screening: A Reply to Swint and Kaback', Stuart F. Spicker has two primary concerns. First, he is critical of the balancing or trade-off metaphor that forms the basis for the authors' discussion. Second, he questions the authors' treatment of the abortion issue. Spicker argues that the notion that genetic screening policy involves balancing economic efficiency with ethical principles is not helpful. He contends that such a view seems to imply objectively determinable criteria which simply must be scaled properly to reach a solution. This characterization is misleading, since it fails to recognize the subjective nature of these criteria. Hence, the notion of balancing raises more questions than it provides solutions.

He also contends that the authors fail to deal with the centrality of the abortion question in the debate over these screening programs. According to Spicker, the essential question is determined by comparing the costs (both moral and economic) of terminating the pregnancy by voluntary abortion. If no one intended to abort affected fetuses, screening programs would have to be justified for purposes other than those typically mentioned.

While the objective of equal access to an adequate level of health care has many adherents, there is considerable disagreement both regarding how the concept is to be justified and how it is to be operationalized. In 'Rationing Medical Care: Processes for Defining Adequacy', Mary Ann Baily primarily addresses the second of these problems. She points out there are three difficulties associated with developing an operational specification of the decent or basic minimum concept. First, there is no consensus regarding the content of a decent minimum, because the care presently available to those who cannot afford to pay varies arbitrarily along dimensions unrelated to the health condition in question and because it is unclear how a principled consensus regarding the basic minimum can be developed in a pluralistic society such as that of the United States. Second, the basic minimum approach seems to conflict with physicians' current perceptions of their ethical duty. Third, there is no consensus as to how much patients must be told regarding the limits on care imposed to conserve resources. Baily discusses these barriers to the implementation of a decent minimum and argues that they are surmountable obstacles.

Baily's approach is to focus on the process whereby the decent minimum approach can be translated into practical public policy rather than to provide a philosophical justification for it. Her position is that the basic minimum approach cannot be a once and for all decision that is imposed on the delivery system. Instead, it must be seen as a process capable of incorporating changes regarding technology, preferences, and resource availability. Similarly, because adequacy is dependent on health conditions, the process requires the active cooperation of providers; and the process utilized must be political, because this is the obvious way to incorporate societal preferences in the United States.

Baily's operational suggestion is to define adequacy as a standard of care acceptable to the middle class. Baily acknowledges that the standard of care *currently* received by the middle class cannot be extended to all. She notes that there is widespread agreement that this standard is wasteful and inefficient. In effect, Baily argues that if the practical problem of the potentially high cost of such a standard could be solved, then there would be support for defining a standard of adequacy that is acceptable to the average person.

She considers two examples of processes for defining and delivering an adequate level of care: the National Health Service (NHS) of Great Britain and a system of competing health care plans as envisaged in Alain Enthoven's Consumer Choice Health Plan ([1], [5]). Although these delivery systems are ideologically opposed (socialized medicine versus market competition), they are similar in the way that each provides for a process which defines and guarantees universal access to a standard of care which approximates adequacy. In both cases, the process is dynamic and various layers of decision making are involved. There is a complex combination of political, administrative, and individual decision-making processes, even though the balance of these is different in the two systems. It is important to note, Baily argues, that in both systems physicians play a key role in setting the necessary limits on the care patients receive. A key justification for this function is the underlying consensus regarding the importance of maintaining a level of adequacy that meets the expectations of a broad plurality of citizens.

David D. Friedman begins his 'Comments on "Rationing Medical Care: Processes for Defining Adequacy"' by challenging the concept of a decent minimum of health care. He argues that the underlying assumption for this proposition that health care is in some sense a special good does not hold up under careful examination. His argument is in two parts. First, he considers the evidence that people do not regard their own lives as infinitely valuable. With respect to the question of whether they should, even if they do not, he points out that a society built on that principle would be one in which all resources were devoted to maximizing life expectancy. Second, for those who believe medical care is special because of its connection to life, he argues that the same connection exists for a wide range of other things. Indeed, one of the main causes of increased life expectancy appears to have been increases in real income.

Friedman goes on to consider the practical questions of whether a standard of care acceptable to the middle class could be provided at a reasonable cost, and if so how. He maintains that such a system is not likely to take on the characteristics suggested by Baily. In discussing the examples discussed by Baily, he notes that neither the NHS nor the Consumer Choice Health Plan avoids some of the faults of our present system in providing a decent minimum to all.

In 'Rationing and Publicity', Gerald R. Winslow addresses a corollary problem of rationing, namely whether the standards should be publicized or kept secret. Winslow reflects on one of the moral "costs" of losing innocence as society is forced to face explicitly the fact that some of the lives that could be saved by medical care are simply not worth saving. What is unique about the present situation is that the capacity to ascertain the cost-worthiness of care is growing and, as a result, limitations that were implicit earlier are now becoming open to view. Indeed, as the concept of the decent minimum continues to gain acceptance, rationing becomes a permanent and officially sanctioned feature of the health care system. Winslow's question simply is: Should the rules for rationing lifesaving medical care be made public?

Winslow is primarily concerned with macroallocational decisions — that is, decisions regarding which classes of patients should be provided with what kinds of medical care — but his view has obvious implications for microallocational decisions as well. The case for publicity is advanced by Winslow in terms of six considerations: (1) Publicity discourages the institution of inequitable rules that are incapable of universalization; (2) publicity enhances the stability of cooperative morality by enlisting the support of society's members; (3) publicity is necessary in order for moral rules to be taught; (4) publicity aids the establishment of valid, mutual expectation; (5) publicity assists those who need to appeal injustices; and (6) publicity promotes the criticism of rules that are in need of change. Underlying these reasons is the view that publicity is required if persons are to be fully respected as autonomous beings.

Winslow also considers objections to publicity that turn on the view that in order to achieve certain justified and necessary social objectives non-publicity — and even deception — is required. In this regard, Winslow discusses the views of Henry Sidgwick, Steven Rhoads, and Guido Calabresi and Philip Bobbitt ([12], [10], [4]). In a concluding section, Winslow answers the objections to publicity offered by these authors. He argues that predictions of the dire effects of publicity are difficult to assess. Winslow agrees that the costs of publicity for rationing would be significant, but notes that they are necessary in a democratic society. Although publicizing approaches to rationing might produce inefficient changes in macroallocation, it is far from clear that this necessarily is the case. In fact, publicity of the inefficiences is essential to their eventual correction. Thus, Winslow concludes that publicizing the rules for rationing, although possibly costly, is necessary.

In considering Winslow's points on the need for publicizing rules for rationing in 'Comments on "Rationing and Publicity"', David D. Friedman makes the important distinction between "fundamental" rules that define entitlements and rules that determine how medical care is allocated. He argues that rules that define and allocate property define who owns what, what an owner can do with what he owns, how ownership is transferred, etc. These rules must be public so that all of the participants may know what they can or cannot legally do. However, the rules that Winslow refers to are related to the preferences of the numerous individuals that control the relevant resources. Friedman finds little support for an obligation on the part of all of those individuals to state publicly the grounds on which they are acting.

The concluding section, entitled 'Controlling Costs/Maximizing Profits: The Role of Providers', addresses some conflicts which providers face in the new cost-conscious atmosphere. In general, two approaches to the problem of cost-containment have been evident in the United States: rationing by government, private payers, and health care plans through administrative rules and rationing by providers and patients in response to economic incentives. In 'Physicians and Cost Control', Charles E. Begley considers some of the implications associated with the rationing of medical care by providers and patients in response to economic incentives. Many commentators are concerned regarding the ethical ramifications of these changes. In his essay, Begley discusses three issues which make assessing the ethical dilemmas faced by physicians particularly difficult. First, because of uncertainties regarding the benefits of medical care, it is difficult to know whether cost considerations pose a serious threat to the quality of care. Second, where conflicts arise, it is not clear whether the physician or patient perspective should prevail. The doctrine of informed consent gives primacy to patient values, yet distributive justice considerations suggest that individual patients' choice of care may be overridden by societal concerns in some instances. Third, conflicts regarding legal standards of care and informed consent complicate efforts to introduce cost control into medical practice. Such standards hold physicians to a fiduciary relationship which is predicated primarily on the well-being of the individual patient without regard for alternative uses of resources or other social perspectives.

In 'Shifting Priorities and Values: A Challenge to the Hospital's Mission', Marc D. Hiller and Robin D. Gorsky suggest that discussion regarding the introduction of cost-containment, procompetition-based models of care, and the corporatization of medical practice in the United States has tended to focus either on the societal level where questions of access to medical care are primarily addressed or on the individual patient care level involving obligations within the physician-patient relationship. They shift attention to what they term the messo level of the health care institution, particularly the hospital, as an instructive focus for discussion. They argue that the paradigm dilemma arising at this level is the conflict between a traditional charitable healing mission of the hospital, which involves benefitting patients without limitation of cost or other factors, and the new business ethic, which is geared towards maximizing institutional profitability.

Hiller and Gorsky argue that hospitals should be regarded as moral agents. Within hospitals, they see the hospital administrator as the pivotal player in addressing the ethical issues caused by the changes in the structure and function of hospitals. They discuss the development of the American hospital in historical perspective and contend that current developments threaten the traditional charitable mission of hospitals. In particular, they view the growth of for-profit health care institutions with alarm and articulate worries regarding limitations of access and quality of care that such changes seem to portend. Much discussed strategies, such as refusing admission to uninsured, underinsured, or financially destitute patients, transferring the indigent to public institutions, and early discharge, are seen by the authors as representing a serious and significant departure from the traditional healing mission of the hospital.

Lu Ann Aday, however, points out in 'Shifting The Question About the Hospital's Mission' that these concerns, though legitimate, are in fact hypotheses for which there is currently very little empirical data. To be sure, there is some ground for the fear that the effort to utilize fewer resources in order to maximize profit will necessarily result in poorer care, but available evidence suggests that more care is not necessarily better. Both Aday and Hiller and Gorsky, however, agree that analysis and discussion of current health policy issues should be understood in an historical context. They differ, however, precisely on the proper interpretation of the development of the historical role of the hospital and the conclusions which can be drawn therefrom.

Aday offers a less pessimistic interpretation than Hiller and Gorsky of the changes currently affecting the structure of health care in America. In the first place, Aday points out that Hiller and Gorsky's account stresses the watershed character of the changes currently under way, whereas a more balanced view would acknowledge that American medicine has long included significant for-profit components and corporate medical practice. It is not the rise of for-profit institutions as such that is most important, but rather the purchase of hospitals by investor-owned corporations to form for-profit multihospital systems. The growth of these corporate entities is a strikingly new development in American medicine, but it is premature, Aday argues, to draw conclusions regarding their impact on access and quality of care.

If hospitals consistently functioned as purely charitable institutions never realizing a profit, but regularly sustaining losses, Aday points out such institutions would quickly cease to exist. It is not profits as such that are new or problematic in hospital care, but rather the possible diversion of profits away from the capital base for health care. Similarly, Aday stresses that Hiller and Gorsky, in focusing on the institution and hospital administrators, fail to consider the contribution of the community served by the hospital in setting hospital policy and style of practice. She cites literature which stresses the role of boards of trustees rather than administrators as playing the key role in this regard.

As the cost-containment effort intensifies, its conflict or agreement with other competing concerns and objectives must be made explicit. To do so requires information regarding the major normative values involved in applying economic criteria to medical decisions as well as how these criteria compare to other standards or norms which guide medical decision making. The authors in this volume attempt to provide such information by examining various questions concerned with the ethical and philosophical implications of the economic analysis of health care.

GEORGE J. AGICH
CHARLES E. BEGLEY

BIBLIOGRAPHY

[1] Aaron, H. J. and Schwartz, W. B.: 1984, *The Painful Prescription: Rationing Hospital Care*, The Brookings Institution, Washington, D.C.

[2] Arras, J. D.: 1982, 'Health Care Vouchers and the Rhetoric of Equity', *Hastings Center Report 11*, 29—39.

[3] Bunker, J. P.: 1985, 'When Doctors Disagree', *New York Review of Books 32*, 7—12.

[4] Calabresi, G. and Bobbitt, P.: 1978, *Tragic Choices: The Conflicts Society Confronts in the Allocation of Tragically Scarce Resources*, W. W. Norton and Company, New York.

[5] Enthoven, A. C.: 1980, *Health Plan: The Only Practical Solution to the Soaring Cost of Medical Care*, Addison-Wesley Publishing Company, Menlo Park, California.

[6] Gibson, R. M., *et al.*: 1984, 'National Health Expenditures, 1983', *Health Care Financing Review 6*, 1—29.

[7] Hicks, J.: 1940, 'The Valuation of Social Income', *Economica New Series 1*, 105—124.

[8] Kaldor, N.: 1939, 'Welfare Propositions of Economic and Interpersonal Comparisons of Utility', *Economic Journal 49*, 549—552.

[9] Klarman, H. E.: 1982, 'The Road to Cost-Effectiveness Analysis', *Milbank Memorial Fund Quarterly/Health and Society 60*, 585—603.

[10] Rhoads, S.: 1980, 'How Much Should We Spend to Save a Life?' in S. Rhoads (ed.), *Valuing Life: Public Policy Dilemmas*, Westview Press, Boulder, Colorado.

[11] Schwartz, W. B.: 1984, 'The Most Painful Prescription', *Newsweek 104*, 24.

[12] Sidgwick, H.: 1966, *The Methods of Ethics*, Dover Publications, New York.

[13] Starr, P.: 1982, *The Social Transformation of American Medicine*, Basic Books, Inc., Publishers, New York.

PART I

MEDICAL ECONOMICS AND ETHICS: SOME THEORETICAL CONSIDERATIONS

MARC J. ROBERTS

ECONOMICS AND THE ALLOCATION OF RESOURCES TO IMPROVE HEALTH

I. INTRODUCTION

This paper provides a review and critique of various ways in which economists have looked at the problem of allocating resources to health care. Three different perspectives that various economists have used at various times are considered: the so-called human capital view, that version of utilitarianism which leads to willingness to pay as the basis for life valuing, and cost-effectiveness analysis. My purpose is two-fold. First, I want to show that although all these ideas have been advanced within the same disciplinary tradition, they have quite different philosophical roots and justifications. They rely on and/or imply quite different views of the relationship between individuals and the state, and of the objectives at which the state is directed.

My second goal, however, goes beyond elucidating the fine points of doctrinal schisms among the economics faithful. Instead, I want to pursue the process Bernard Williams has called "reflective criticism". This involves exploring whether our avowed ideas are or are not consistent with our actual practice, and where these conflict, trying to decide which to accept.[1] In the health care context this enterprise involves examining to what extent the various economic theories are or are not consistent with the ways in which health care resources are actually allocated in the United States. I argue that in fact there has been a sharp conflict between these theories and that practice. This inconsistency, in turn, is subject to two rather divergent interpretations. On the one hand, some economists offer it as proof of the "irrationality" or "inefficiency" of current practice. Alternatively, one can argue that there are some fundamental flaws in or limitations of the economic viewpoint which leave it unable to cope with pervasive and persuasive ethical beliefs.

Which conclusion is valid? Answering that question is a central task confronting those who want to think seriously about the ethical problems of allocating health care resources in the United States today. Exactly because all economic arguments focus on consequences, they

G. J. Agich and C. E. Begley (eds.), The Price of Health, 3–22.

leave out those elements of our decision making that ignore consequences. As a society, we often do appear to do things because they seem to us to be "fair" or "appropriate" regardless of the outcomes they might produce. On the other hand, the most fundamental ethical insight that economics teaches is also relevant – namely, that in a world of limited resources it is imprudent and ultimately indefensible to ignore consequeces altogether when deciding how to proceed.

This tension, I will contend, can, should, and no doubt will be resolved by political processes. My task is to pose some questions clearly enough so that they can become a fruitful focus of public discussion and deliberation. I will close by arguing that the most illuminating and constructive way to formulate these questions is to avoid the language of "rights" and entitlements. We would do much better by asking ourselves what kind of society do we want to live in? What kinds of support do we want to provide for each other by virtue of our common citizenship and our common humanity? Only in that way can we begin to reconcile efficiency and generosity, charity and citizenship, fairness and prudence as we struggle to allocate health care resources.

II. THREE ECONOMIC IDEAS

From the earliest days economists have had to confront a deep dualism between descriptive and prescriptive ideas that runs through the conceptual system of their discipline. Some practitioners have seen themselves self-consciously as the imitators and emulators of Newton – impartial observers trying to write down the social equivalent of the laws of planetary motion ([6], [42]). Yet, interest in economic (unlike astronomical) process has almost always been driven by practical as opposed to aesthetic or cognitive concerns. Economists have repeatedly been asked for, or on their own accord have offered, ideas about social policy.

The question of whether such suggestions were and are "scientific" has been a continually vexing one within the profession. Is it possible to determine what is "efficient" or "economically rational" without invoking some set or other of disputable ethical assumptions? Most students of political philosophy would agree that even to decide that efficiency is worth studying one must utilize some ethical assumptions to justify that conclusion. Characterizing an outcome as "efficient" is only interesting if we care about "efficiency" defined in that way. Deciding what to care about is not a scientific question. In contrast, many economists, especially today, seem to be much less aware of and self-conscious about the ethical presumptions built into their conceptual systems.[2]

As a descriptive endeavor, economics has been concerned with how and why the resources available to a society are used as they are to produce the outputs that are available. As a prescriptive endeavor, economists have been concerned to say how those limited, scarce resources should be employed. The problem economists face is that the economy produces many different things. In order to compare outcomes, one needs to have a way to add up or aggregate these diverse outputs on the basis of some index of desirability.

The value-based nature of economic analysis comes across clearly when we consider that different measures of desirability have been offered, each with different ethical underpinnings. The three major views that economists have offered about how to allocate health care resources sharply illustrate this point. The oldest and in practice the most widely used measure of the value of alternative health promoting practices is the so-called "human capital" method [7]. This technique values lifesaving (or by extension, morbidity avoided) by asking how much the individuals in question would have produced if they were left alone and/or unimpaired. A second, and much more recent, notion is that the value of health improvements (like the value of anything else) can be determined by what people would be willing to pay for the "output" in question ([37], [26]). Such a question, presumably, is to be asked not only of the individual concerned but also of everyone else concerned with his health. While both of these methods provide a direct monetary value to lifesaving, the third method, cost-effectiveness analysis, is more modest in its claim. It merely proposes to insure that whatever resources spent to improve health are spent to get the maximum possible result [44]. Insuring this, in turn, requires that we have some way to measure, compare, and add up "health" consequences of various kinds as they occur to various individuals.

Human capital measures of life valuing owe their origins less to economists than to public health professionals and early "vital statisticians". As early as the seventeenth century, Sir William Petty [29] was using this approach, and it was later adopted as well by William Farr [10]. These methods were used to argue for various policy proposals, from the economic gains to be had from public health investments to the virtues of allowing skilled workers to emigrate into England from abroad, as well as to analyze the costs of war. By the close of the nineteenth century, this approach had the blessing of Alfred Marshall, the pre-eminent English economist of his day, who analyzed the rational and irrational motives that lead families and employers to invest in this human capital [23]. He used the general idea as a significant part of his overall discussion of patterns of labor supply and income.

In America, Yale economist Irving Fisher used the same approach while also developing the theoretical arguments as to why future earnings should be "discounted" — that is, valued less the further in the future they occur based on the rate of interest [13]. While there was some very interesting work on this question done in the 1930s, particularly an intriguing and thorough book by L. I. Dublin and A. J. Lotka [7], the real blossoming of interest came in the post-World War II period. Again, it was a "vital statistician", Dorothy Rice, rather than an economist, who took the lead and computed extensive tables of average life values by age, race, and sex ([31], [32]).

The virtues of the numbers derived by this method included the practicality of actually making such computations (with the aid of many simplifying assumptions) on the basis of available income data. And they were easily characterized as "scientific" and "rational" by their proponents. Difficulties, however, were also apparent. Since the time of Marshall, some economists had recognized that some "services" were "unpriced". Hence, those who produced such services (especially women working at home) would have a low life-value as seen by these methods ([7], p. 15). Similarly, any discrimination or other imperfections in the labor market would also affect these numbers. Most fundamentally, however, was the question of whether the value of a life, and hence of lifesaving, could or should be calculated solely on the basis of the output that life would produce. After all, why should a society maximize the discounted net present value of its Gross National Product — which is the result toward which this decision rule was directed?[3]

To put it another way, what kind of social theory could justify using this life-valuing technique? On reflection, the whole idea seems squarely in the tradition of Plato and the medieval thinkers who viewed the state as an organic entity with interests that each member is supposed to serve. Such an organic state is likely to be ruled by experts of one sort or another who know how to achieve those ends. They might be the philosopher king of the Republic, Burkian aristocrats, Saint-Simonian engineers, the Leninist vanguard of the proletariat, or the Dalai Lama and his aides – depending on the particular ends to be sought and the correlative expertise that is deemed relevant.[4]

While non-economists tended to view the human capital method as the "economic" approach to priority setting, the main line of economic thinking was taking a different direction. Economists since Adam Smith have approached the question from the perspective of the individual whose life was being valued. They began by recognizing that the non-monetary aspects of a job such as amenity, danger, difficulty, and status affected the employment decisions of workers and the compensation they demanded. From this observation it was only a small step to the notion that the costs of danger, and hence the value of life and lifesaving, were being judged by workers themselves everyday in the marketplace. This value would be reflected in the risk premiums that were required to induce workers into more hazardous occupations [39].

This analysis intersected more general developments in what has come to be called "welfare economics" [25]. The problem of efficiently allocating resources is after all not unique to the arena of health. The main line of economic thinking on this question has been utilitarian. The output of the economy was to be judged by how happy it made various individuals. The problem has always been that most real economic policies help some and hurt others. Since Jeremy Bentham economists have struggled with the question of how to measure gains to gainers and losses to losers so that we can add them [2]. This has become known as the problem of "interpersonal comparisons of utility" [20].

While some writers in America between the world wars, notably Irving Fisher, seemed to believe (mistakenly) that they had such a method [12], the predominant view in Europe was that of Vilfredo Pareto, namely that such comparisons were impossible [28]. This led Pareto to propose that economists could only approve of policies from which everyone gained. Unfortunately, since such policies were rare indeed, this unanimity rule would have reduced economic analysis to irrelevance. Two economists working in England, Sir John Hicks and Nicholas Kaldor, proposed what came to be called "the Kaldor-Hicks test" ([16], [17]; see also [15], [20], [25]). That is, a policy was to be approved not only if it *did* make everyone better off, but if it *could* make everyone better off — even if, in fact, it did not do so. This involved asking if the "gainers" from some proposal could *potentially* "compensate" the "losers" and still have something left over for themselves.

In the immediate post-World War II period, this suggestion came under severe attack from the most sophisticated theoretical economists. Tibor Scitovsky [37] and Paul Samuelson [35] pointed out first that the test could be circular. Point A could seem better than B, and B then seem better than A. More fundamentally, Samuelson [36] and Graaff [15] argued that the test assumed the appropriateness of one or another income distribution together with the resulting set of prices, production levels, etc. Hence, a set of arrangements that looked satisfactory from one perspective could look unsatisfactory from another. (The possible intransitivity of the criteria was just a reflection of this point.)

In *The Foundations of Economic Analysis,* Samuelson ([35], pp. 249—52) proclaimed the end of the Kaldor-Hicks method as confusing and confused and instead argued for the need to evaluate all changes in terms of an overall "social welfare function", an idea first proposed in the 1930s by Abram Bergson [3]. Individuals' relative utility levels — as revealed by their individual choices — were to be combined in an overall ranking of social states.

However, despite its theoretical limitations, the Kaldor-Hicks idea has survived in the form of modern cost-benefit analysis. That method involves measuring the willingness to pay of gainers for their gains and comparing these with costs. If all markets are assumed to be "perfect", those costs can be shown to be equal to what other consumers would have paid for the outputs that are now not produced because resources have been diverted to the activity in question. The idea actually has the roots in still earlier work by the French engineer, Jules Dupuit, who in the nineteenth century struggled with the problem of deciding whether or not to build a bridge over the Seine that would be used for free, and devised the same method.[5]

Despite its roots in utilitarianism, the method as practiced cannot easily be defended on those terms. "Dollars" do not translate into "utilities" in any simple, interpersonally comparable fashion. Similarly, it is not easy to defend the current technique as a sequential strategy for reaching more efficient production. Since R. G. Lipsey and K. Lancaster's famous paper on "the second best" [19], which echoed earlier and less well-known work by Samuelson [34] and I. M. D. Little ([21], pp. xiii–xiv), we have known that piecemeal application of policies designed to produce some of the preconditions for efficiency can in fact make matters worse as long as other imperfections persist in the system.

On reflection, however, it has become clear that the Bergson-Samuelson attempt to modernize utilitarianism was a philosophically naive dead end. The practical problems raised by "interpersonal comparisons of utility" are in fact insurmountable. Human experience is irreducibly multidimensional. Actual "utility levels" as such do not exist; hence, they cannot be measured and compared. It is true, as Samuelson argued, that without such knowledge, we can still (as a logical matter) write down a rule that tells us whether society is to be considered better off when one person moves from circumstance A to circumstance B, which he prefers, while another moves from X to Y, which he does not prefer. But without knowing anything of the psychic reality behind such changes, we cannot defend the rule over the infinity of other possible rankings that might be offered.[6]

In contrast to the elaborate social welfare function approach of Bergson and Samuelson, the advantage of willingness to pay has always been its practicality even though it is not as easy to use as human capital methods. Initial applications in the United States occurred in the water resource area. The Army Corps of Engineers was required by the Water Resource Act of 1936 to build projects for which "the benefits exceed the costs" [31]. In fact, this legislative language also accounts, I believe, for the current terminology, which characterizes such ideas as "cost-benefit analysis". Since the 1960s and 1970s, it has been increasingly applied to everything from tax policy to health policy.

On the other side, in contrast to "human capital", economists generally see willingness to pay as superior because it offers some connection to the broad utilitarian framework most economists have been trained to accept. In such a framework individual health status is an output, not an input. Still, the essential problem is to allocate resources efficiently to maximize output valued in this slightly different way. In philosophical terms, the argument is resolutely "consequentialist", focused as it is on consequences as the test of desirability.

For many engaged in actual resource allocation decisions related to health, however, the method still seems too impractical to be taken seriously. Using risk premiums in wage rates is subject to the objection that real labor markets are highly imperfect, workers ill-informed and not fully free to choose. There is the additional problem that the value of a given life to others presumably can only be elicited by time-consuming and inaccurate questionnaires. More fundamentally, many object to the essential point of the whole approach, which is to value different lives differently depending on the views of the individuals involved. The fact that income and wealth varies greatly and inevitably constrains real willingness to pay also makes many observers dissatisfied with the technique [11].

This has led some observers — both public health professionals and economists — to fall back on organic views of the state. Now, however, unlike in the human capital method, the goal to be maximized is not the G.N.P., but the health status of the population. Within public health this approach has been embodied in the development of "health status indicators" ([14], [9], [24]). Numerical scales for evaluating an individual's health status have been made into priority-setting schemes by adding the values across individuals to produce a population-based value. Economists have come to substantially the same approach when they propose to measure, and add across individuals, so-called "quality adjusted life years," and use those totals as a guide to resource allocation [44].

I characterize these methods as reflecting an "organic" view of the state, because some expert has to make the decisions that define the index. For example, how do we combine or weigh freedom from pain and mental clarity? Almost no one uses such methods to suggest that claims on resources devoted to improving health are unlimited. On the contrary, such concepts are invariably appealed to when the question is how to best allocate a limited budget. Yet that budget constraint (like the weights that define the index) has to come from somewhere. Once we presume that some experts know what the objectives of the state are without consulting its citizens (indeed, even that it has such objectives), we are operating in the organic tradition. Only now we have doctors and those trained in "clinical decision theory" as the experts to tell us how to be "efficient" in terms of this new presumed social goal. The whole approach is clearly seen for what it is when we realize it is just like wartime triage only transposed and generalized to a peacetime setting.

III. CONFLICTS WITH ECONOMICS

While all economic ideas assume slightly different objectives, they are united in their commitment to use resources efficiently. Yet much health care practice in advanced countries is hard to justify in terms of any of these notions. Instead, such practice is filled with apparent contradictions and inconsistencies. Consider the following examples.

— Substantial resources are now often spent to improve the living conditions of severely handicapped individuals who are in some cases bedridden and barely able to function. Neither health, utility, nor economic productivity is likely to be much increased by some of these expenditures, especially on the margin.

— There are documented instances of expensive surgical procedures that are routinely undertaken in emergency situations that do little or no good in the sense of improving health outcomes [27]. Such expenditures are hard to justify from any consequentialist perspective.

— We observe wide variations in patterns of practice among regions in terms of surgery rates, lengths of stay, ancillary utilization, etc. [44]. Obviously, resources are not being allocated to consistently advance one formulation of our objectives across all these regions.

— While some health care is sold through the marketplace (for example, much cosmetic surgery), we have aggressively provided public financing to remove other procedures (for example, kidney dialysis) from much willingness to pay constraints [33].

— We provide certain levels of medical care even to those persons sentenced to death. This is hardly a very promising investment from either a GNP maximization or health maximization point of view.

— We provide people with incentives to smoke and to drive badly by publicly financing much of their health care if they become unwell as a result of these activities, even though we also undertake various policies to discourage such behavior.

— We typically find various ways to cross-subsidize and to have others implicitly finance the care of those who are uninsured — even when an individual's insurance status is of his own doing.

In fact, many of these expenditues are argued for or justified by claims that fly in the face of all three of the economic perspectives just reviewed. That is, they are often supported by appeals to justice, charity, or most frequently, rights of one sort or another. These arguments are assiduously "non-consequentialist" in form or, as philosophers would say, they are "deontological", deriving their force from claims based "on the nature of things" rather than by identifying the results that various actions are likely to achieve. In order to consider how, if at all, such arguments can be reconciled with economic ideas, I want to explore a bit further how such claims might be established and of what they might consist.

IV. VARIOUS CLAIMS TO HEALTH IMPROVING RESOURCES

In fact, there are a wide variety of claims made about how resources that might improve health "ought" to be allocated. The first distinction we have to make among these various kinds of claims has to do with the content of the claim: to what exactly does the claim apply. The relevant distinction here is between inputs and outputs, between health care and health.

Oddly enough, there are several reasons for arguing that fully non-consequentialist claims — that is, claims to resources and not to outcomes — are easier to reconcile with or integrate with consequentialist ideas. By hypothesis, a limited claim to some resources — even if it is made on non-consequentialist grounds — is still potentially consistent with some concern for outcomes both in the use of those resources and in the use of other resources for other purposes. We can say "citizens ought to have certain care, regardless of the results", and still try to do our best to achieve our other objectives, as economists would say, "subject to this constraint".

In contrast, a claim to a certain health status outcome raises far more difficult problems. While such claims are not that common in the philosophical literature, advocates who insist that some class of patients "ought" to have access to all the care they "need" are in fact making such arguments. Yet when we think about the matter carefully, the reality of death makes the claim to "health" ultimately unfulfillable and hence potentially unlimited. Indeed, death aside, an individual's "health" is in fact a highly complex, multidimensional, variable state. There is no one well-defined outcome that constitutes health. And even if some level of functioning and physiology were arbitrarily chosen as the benchmark, it would not always be possible to achieve that result in any given patient. Again, claims are ambiguous and arguably unlimited. Furthermore, there may often be situations in which small improvements in health can only be bought at great price. Finally, asserting a claim to health against society ignores or turns our attention away from the major role any individual's own behavior plays in determining his health status.[7] Do we as a society really want to insure people fully against the adverse health consequences of smoking, drinking, poor diet, bad driving, committing crimes, taking great risks for sport and so on?

These points lend support to the view that non-consequentialist claims can most usefully be thought of as claims to care, not to cure. That, to be sure, still leaves many vexatious questions unanswered: To what extent or in what way does the care to be provided depend on the likelihood that this patient in general, or all patients on average, will benefit from it? For example, should we perform expensive procedures like dialysis and transplantation on those likely to die soon of other causes? Similarly, should the level of care provided vary with the wealth, circumstances and/or traditions of a particular society? If there are "rights" to care, are they the same in the United States and Canada, or the same in the United States today as they were ten years ago?

A related question is whether, in a federal system, the nation is the relevant unit for making such decisions. Instead, might such decisions fall, at least in part, to subordinate jurisdictions? With state operation of Medicaid and the local government role in public hospital operations, some decentralization is certainly characteristic of the United States today.

Similarly, does the care provided depend on the patients' role in producing their own condition? Do we offer, at public expense, lung surgery for cancer victims who smoke, trauma care for injured drunk drivers, liver transplants for alcoholics, or AIDS treatment for intravenous drug users and homosexuals who contracted the disease once its transmission routes were well known? If we are going to provide care for a condition, just what does that entail or involve? Does it mean any and every possible resource? Does it mean any and every amenity? If the middle class chooses "Volvo medicine" for itself, do we have some civic obligation to provide similar amenity and effectiveness to all citizens or will the medical equivalent of a basic Chevrolet without air conditioning suffice?

V. TOWARD A SYNTHESIS

Resolving these various questions is not an easy task. How people believe we should proceed is likely to depend on their particular view of the appropriate relationship of individuals to society and their views on the nature of moral argument and inquiry. Here, I can do no more than sketch a very brief and incomplete account of my own position — as incoherent and inchoate as it is — and use that as a way of indicating possible future directions of political and academic effort. I do believe that in this realm the two are likely to be deeply interconnected.

As Alasdair MacIntyre has rightly observed, much of the history of modern moral philosophy has consisted of trying to find a method or standard for justifying moral rules that was at once both universal and not supernatural [22]. The desire to claim universality, in particular, has imposed a heavy methodological cost. It meant the need to presume, as Kant understood, that moral ideas must somehow be knowable *a priori*; otherwise, they would be conditional on particular experience and hence not universal.

It now seems increasingly evident that this effort to ground morality on reason has failed, and failed because it set itself an impossible task. Reason that is not based on experience can deal only with the materials that reason by itself can create — that is, with closed postulational systems. Yet, before we can apply arguments within such a system to real situations, we have to decide whether or not to accept the relevant postulates. That, however, is not a task reason, by itself, can adequately perform. Just as we cannot go from the descriptive to the normative, so too we cannot leap over the logical chasm between the formal and the normative unless we have the aid of non-formal, substantive, ethical ideas. If such ideas can come from neither God nor reason, then they must come from human experience. But what kind of experience? The experience of purely observational science is insufficient since it alone cannot lead to ethical imperatives.

One emerging line of thought, visible in the work of Michael Sandel [37], Michael Walzer [42], and Bernard Williams [45], is that the non-universal experience of ourselves and our culture both does and must serve as the ultimate reference point. The question is not what should all men and women believe, but what in fact do certain individuals (namely, we ourselves) believe to be morally right.

The advocates of this perspective have had to fight a philosophical war on several fronts. For example, Sandel has criticized John Rawls' attempt to ground morality in "fairness" and to ground "fairness" in the choices hypothetical individuals would make under certain hypothetical circumstances. He has argued that individuals stripped of all history and all personal characteristics — as Rawls would have us assume — would not be able to choose social rules exactly because they would have no basis for doing so. Similarly, Williams has criticized utilitarianism on the grounds that it fails to take account of many of the complexities of moral thought as accepted in the society. And Waltzer has argued, against the libertarians, that the community, not the individual, is the basic unit for analyzing distributive moral issues.

How, then, are we to decide what claims, if any, to accept as constraints on the purposeful allocation of health care resources? Indeed, how do we as a society decide what purposes, if any, such resource allocation decisions should serve? The common thread in all these "communitarian" arguments is that decisions about what efforts society should make to improve the health of its members is, by its nature, an essentially collective, civic decision. As Sandel has put it, it must be grounded in a "thick" sense of self, one rooted in specific historic experience and social commitments. Thus, the relevant question to ask is what goals, rights, obligations, or entitlements does the traditions and experience of the community suggest should be accepted or established. If these are community matters, then how is the particular community defined at this time? What does membership in it entail and imply in the health arena? What can we expect and what should we work for?

As I and others have argued elsewhere [18], the processes that might most fruitfully explore such a question are both civic and deliberative. The complexity and ambiguity of our own moral ideas means that each of us has much to learn about their meaning and implications in particular circumstances. Conversation with others, based on some minimum sense of shared values and common purposes, is one way to foster such an understanding. This is particularly so when question requires individuals to meld both self-regarding and other-regarding perspectives. The issue is not simply what each of us as individuals wants. Rather, what does each of us believe is a fair and appropriate expression of our mutual obligations by virtue of our common citizenship in a particular political community. Given this orientation, we can inquire how our current practices depart from our supposed ideals, and what does that tell us about the deficiencies of either our behavior or our theory? More particularly, since otherwise we would lose much of our capacity to criticize current practice, we must ask: What allocation of health care resources fits with our highest and best view of ourselves and our shared community?

There does seem to be a widespread desire to insure that individuals who could substantially benefit from medical care services do not find it difficult or impossible to obtain such services because of their income, race, or geographic location. This commitment seems quite widespread — as the broad acceptance of the "safety net" metaphor across the political spectrum indicates. The nation does seem willing to provide some basic minimum or "adequate level" of care, to use the phraseology of the President's Commission ([30], p. 20). Furthermore, such a minimum can be argued for in terms of any one of a wide variety of ethical commitments from fairness to charity. All of these have the common theme that the status of individuals as members of a common community creates certain reciprocal rights and obligations. Indeed, anyone who takes notions of citizenship and community seriously would be hard pressed to deny the appropriateness of some level of support for those who could benefit from basic care.

The difficulty arises when we try to define more clearly just what such support should consist of under various circumstances. This problem is particularly vexing in the context of potentially lifesaving services which are both very expensive and often ineffective — especially for the most seriously ill patients. So far society has found it very difficult to limit such care, even when the distinct possibility exists that widespread use of some particular therapy would divert resources from other more effective health care measures. As a result, activities like the End Stage Renal Disease Program have provided publicly funded care for patients who in other societies have been considered too unpromising to treat.

Notice, however, that part of the appeal of the "decent minimum" idea is that it involves only a limited claim, and appears to respect the legitimacy of resources claims for other competing purposes. Thus, those who support this approach have an important obligation to say where those limits are. As advocates who often argue in favor of more care, they have a particular responsibility to focus public attention on the possibility that some limits on care, even to those who are reasonably likely to benefit from it, are consistent with their approach.

Otherwise, we run the risk that the slogan of a "decent minimum" will be used to justify the same potentially limitless resource claims that a more naive claim of "rights to health" would generate.

In that context, I believe there is much to be said for making a planning decision to limit society's capacity to provide certain services. The medical community can allocate the resulting capacity (for example, for organ transplantation) to those who would most benefit. Such resource limits, and the comparative judgements they lead to, have been socially acceptable in the past. If we go the alternative route of financing all "in need", while trying to limit our definition of the "needy", we will face a hopeless task in turning away marginal cases. This is especially difficult to justify since the "craft" element of medical science makes it difficult to know exactly how much any given patient will benefit from a procedure. By the same token, there is some evidence that practitioners' ideas of "appropriateness" adjust to rationalize available resource levels, so that whatever we provide will come to seem about right over time. We can always increase that capacity if experience shows us that we are turning away patients on the margin who would benefit significantly.

If we do not impose such limits the rewards within the health care system will encourage aggressive service expansion since providers are paid more when they provide more care. The result will be that the "decent minimum" will become so costly that political protests will threaten its provision. The delivery system surely will defend its interests in "high tech" activities, leading to cutbacks in basic services like prenatal clinic care. The point is that any definition of the content of the "decent minimum" has to include a specification not only of what services will be available but also to whom. Unless the latter is limited to those who meet some reasonable threshold test of likelihood of significant benefit, the notion loses its power as a *limited* claim.

Another issue society has to address is the role of "fault" as a potential obstacle to obtaining such services. We do not, in general, let injured drunk drivers die when they slam into trees late at night while driving at excessive speeds. Yet, recently we have become increasingly conscious of the role of behavior in producing health status, and have begun to consider a variety of incentives, educational and regulatory efforts directed at such parameters — for example, warnings on cigarette packages, educational programs aimed at teenage pregnancy, and tougher speed law enforcement.

To what extent are we willing to continue current practice with regard to self-induced harms? My own view is that however much we as a society try to use education and incentives before the fact to get people to change their behavior, we are unlikely to refuse to provide them with some minimum of care even when they do not respond to our efforts. And it is not clear to me that the adverse incentive effects of such expressions of charity and social solidarity are large enough to compel us, as a society, to abandon such individuals.

Finally, we must realize that this debate is occurring, and will occur, at a time when the enthusiasm for government action and the general social optimism characteristic of the 1960s is steadily diminishing under the pressure of changing worldwide economic conditions. More foreign competition, slower growth, lesser tax collections, and fiscal stringency at all levels of government have produced a willingness to question old assumptions. Does fairness demand that the poor have exactly the same care as the rich? Ten or fifteen years ago, many more would have answered "yes" than would do so today. This willingness to limit care has been reinforced by a growing skepticism, both inside and outside medicine, of the effectiveness of many therapeutic decisions [5]. Second opinion programs for surgical decisions, the claims of HMOs to have lower costs and no worse health outcomes, and academic studies of patterns of practice have all played a role in this new climate of opinion [41].

Out of all this I detect an increased willingness, even a perceived feeling of necessity, to confront the question of what kinds and level of health care (and prevention) society should provide for its members. And in particular, there is an increased willingness to ask how much of a role should the consequences of care (as weighed by whom?) play in those decisions.

My own recommendation to those who would participate in this emerging debate is that posing the argument in terms of "rights" may not be a particularly helpful way to state the issue. The attraction of such language to policy advocates is exactly the source of its danger — the apparently uncompromising character of the resulting claim. Americans are sensitive about "rights" and do not override them easily. If health care in general, or certain services, can be characterized as a "right", its insulation from competing resource claims is increased.

No doubt many of those academic writers who have explored the "rights" idea have not used such language for the kinds of advocacy purposes just noted. Still, they have to consider whether or not such concepts are the only, or the best way, to describe what is at stake here. The alternative justification is straightforward. Most public services such as police protection or roads are provided because the community considers them "good" without the creation of specific entitlements on the part of those who benefit. The advantage of this formulation is that it calls our attention to the economist's realization that there are more "goods" that we might desire than there are resources to produce them and hence that choices must be made.

The objection to "rights" language is that in a world in which the limitations on resources are increasingly evident, multiplying constraints on resource allocation is a serious step which prudence dictates should only be done cautiously and for compelling reasons. This is particularly true since, as a society, our ability to learn from experience is highly asymmetrical. We can always do more. (It is hard to do less — especially once expectations have been created.)

Resources are scarce. Funds spent for heart transplants could go for food programs for poor mothers or even into general economic development activities — not to mention education, environment, transportation, and defense. It is important to ask what kind of care do we want to be assured of for ourselves and each other. What would a good society provide for its citizens insofar as it is able? What does the ideal of charity, even generosity, contribute to our understanding of the implications of our "civic friendship" ([37] p. 180—81) and where must we, perhaps regretfully, reject such motivations because of the overall situation in which we find ourselves? Clearly, we have a long way to go in constructing an effective public discussion of these matters. But at least we can begin by recognizing their complexity as well as how the ways in which we frame questions serve to determine answers.

Harvard University,
School of Public Health
Cambridge, Massachusetts

NOTES

[1] Williams has offered this as a modification of Rawls' notion of "reflective equilibrium", arguing that the latter is too static and fails to capture the exploratory nature of the process ([45], pp. 112—117).

[2] The extent to which economists are self-confident and unself-critical is illustrated by the following quotation from one of the leading current textbooks on normative economic methods: "Policy choice in the United States would be much improved if it always started by identifying the consequences of choice for individual welfare, since these, as we have stressed, are the critical components of social welfare . . . the welfare of society depends wholly on the welfare of individuals" ([38], pp. 283—286).

[3] The Gross National Product is the measured value of the output of a nation's economy. Maximizing the amount of productive value saved through health care amounts to maximizing the GNP.

[4] The variation in ends and their associated experts illustrates that this class of theories is very broad indeed. However, the general stance or methods they all use for ordering society are clearly parallel.

[5] Dupuit's original paper appeared in French in 1844. It is reprinted in an English translation [8].

[6] While Samuelson has always refused to see this point, other economists have been more understanding.

[7] The three leading causes of mortality in the United States today are cardiovascular disease, neoplasms and trauma. Smoking, drinking, diet, etc., are all important causal factors in these conditions.

BIBLIOGRAPHY

[1] Arrow, K.: 1963, *Social Choice and Individual Values,* 2nd edition, Yale University Press, New York.

[2] Bentham, J.: 'The Psychology of Economic Man', in W. Stark (ed.), *Jeremy Bentham's Economic Writings,* Vol. 3, George Allen and Ltd., London.

[3] Bergson, A.: 1938, 'A Reformulation of Certain Aspects of Welfare Economics', *Quarterly Journal of Economics 52,* 310—334.

[4] Bergson, A.: 1966, 'On Social Welfare Once More', in *Essays in Normative Economics,* Harvard University Press, Cambridge, Massachusetts, pp. 51—90.

[5] Bunker, J. *et al.* (eds.): 1977, *Costs, Risks, and Benefits of Surgery,* Oxford University Press, Oxford.

[6] Cassel, G.: 1935, On *Quantative Thinking in Economics,* Oxford University Press, Oxford.

[7] Dublin, L. I. and Lotka, A. J.: 1930, *The Money Value of a Man,* Ronald Press, New York.

[8] Dupuit, J.: 1952 'On The Measurement of the Utility of Public Works', *International Economic Papers 2,* 83—110.

[9] Fanshel, S. and Bush, J. W.: 1970, 'A Health Status Index and its Application', *Operations Research 18,* 1021—66.

[10] Farr, W.: 1853, 'The Income and Property Tax', *Journal of the Statistical Society of London 16,* 1—44.

[11] Fein, R.: 1972, 'On Achieving Access and Equity in Health Care', *Milbank Memorial Fund Quarterly 50,* 157—190.

[12] Fisher, I.: 1927, 'A Statistical Method for Measuring "Marginal Utility" and Testing the Justice of a Progressive Income Tax', in *Economic Essays in Honor of John Bates Clark*, Macmillan and Co., New York.
[13] Fisher, I.: 1930, *The Theory of Interest*, MacMillan and Co., New York.
[14] Goldsmith, S.: 1972, 'The Status of Health Status Indicators', *Health Services Reports 87*, 213—20.
[15] Graaff, J.: 1957, *Theoretical Welfare Economics*, Cambridge University Press, Cambridge.
[16] Hicks, J.: 1940, 'The Valuation of Social Income', *Economica New Series 7*, 105—124.
[17] Kaldor, N.: 1939, 'Welfare Propositions of Economics and Interpersonal Comparisons of Utility', *Economic Journal 49*, 549—552.
[18] Landy, M., *et al.: Beyond Pluralism: Lessons for Government Policy Making From the Environmental Protection Agency*, unpublished manuscript.
[19] Lipsey, R. G. and Lancaster, K.: 'Welfare Economics and the Theory of the Second Best', *Review of Economic Studies 24*, 11—23.
[20] Little, I. M. D.: 1950, *Welfare Economics*, Oxford University Press, Oxford.
[21] Little, I. M. D.: 1953, *The Prince of Fuel*, Oxford University Press, Oxford.
[22] MacIntyre, A.: 1981, *After Virtue*, University of Notre Dame Press, Notre Dame, Indiana.
[23] Marshall, A.: 1890, *Principles of Economics*, MacMillan and Co., London.
[24] Miller, J. E.: 1970, 'An Indicator to Aid Management in Assigning Program Priorities', *Public Health Reports 85*, 725—31.
[25] Mishan, E. J.: 1965, 'A Survey of Welfare Economics 1939—1959', in *Surveys of Economic Theory 1*, St. Martins, New York, pp. 154—222.
[26] Mishan, E. J.: 1971, 'Evaluation of Life and Limb: A Theoretical Approach', *Journal of Political Economy 79*, 687—705.
[27] O'Donnell, T. F. *et al.*: 1980, 'The Economic Impact of Acute Variceal Bleeding: Cost-Effectiveness Implications for Medical and Surgical Therapy', *Surgery 88*, 693—701.
[28] Pareto, V.: 1897, *Cours d'Economic Politique Professé à la Université Lausanne*, F. Rouge, Lausanne.
[29] Petty, Sir W.: 1699, *Political Arithmetick*, Robert Clavael, London.
[30] President's Commission for the Study of Ethical Problems in Medicine and Biomedical and Behavioral Research: 1983, *Securing Access to Health Care, Volume One: Report*, U.S. Government Printing Office, Washington D.C.
[31] Prest, A. R. and Turvey, R.: 1966, 'Cost Benefit Analysis: A Survey', in *Surveys of Economic Theory 3*, St. Martins Press, New York, pp. 155—207.
[32] Rice, D.: 1967, 'Estimating the Cost of Illness', *American Journal of Public Health 57*, 424—440.
[33] Rice, D. and Cooper, E.: 1967, 'The Economic Value of Human Life', *American Journal of Public Health 57*, 1954—1956.
[34] Roberts, S. D. *et al.*: 1980, 'Cost Effective Care of End Stage Renal Disease: A Billion Dollar Question', *Internal Medicine 92*, 943—48.
[35] Samuelson, P. A.: 1947, *Foundations of Economic Analysis*, Harvard University Press, Cambridge, Massachusetts.

[36] Samuelson, P.: 1950, 'Evaluation of Real National Income', *Oxford Economic Papers, New Series 2,* 1—40.

[37] Sandel, M. J.: 1982, *Liberalism and the Limits of Justice,* Harvard University Press, Cambridge, Massachusetts.

[38] Schelling, P. C.: 1968, 'The Life you Save May be Your Own', in S. B. Chase, Jr. (ed.), *Problems in Public Expenditure Analysis,* Brookings Institution, Washington, D.C., pp. 127—163.

[39] Scitovsky, T.: 1941, 'A Note on Welfare Propositions in Economics', *Review of Economic Studies 9,* 77—88.

[40] Stokey, E. and Zeckhauser, R.: 1978, *A Primer for Policy Analysis,* W. W. Norton, New York.

[41] Thaler and Rosen: 1976, 'The Value of Saving a Life: Evidence from the Labor Market,' in N. Terleckyz (ed.), *NBER Studies in Income and Wealth, Number 40,* Columbia University Press, New York.

[42] Walzer, M.: 1983, *Spheres of Justice,* Basic Books, Inc., New York.

[43] Wennberg, J. and Gittlesohn, A.: 1982, 'Variations in Medical Care Among Small Areas', *Scientific American 246,* 120—134.

[44] Wicksteed, P. H.: 1979, 'The Scope and Method of Political Economy', *Economic Journal 24,* 1—23.

[45] Williams, B: 1985, *Ethics and the Limits of Philosophy,* Harvard University Press, Cambridge.

[46] Zeckhauser, R. and Shepard, D.: 1976, 'Where Now for Saving Lives', *Law and Contemporary Problems 40,* 4—45.

GEORGE J. AGICH

ECONOMIC COST AND MORAL VALUE

In this essay I consider three common objections to the introduction of economic considerations in medical practice. The objections can be stated straightforwardly. First, the ethics of the physician-patient relationship obligates physicians primarily to do what is right and good for particular patients; it precludes consideration of economic cost or social utility. As a result, the systematic inclusion of efficiency and cost containment in medicine is ethically prohibited, because it would require physicians to participate in the rationing of health care and to make clinical decisions on a basis other than what is right and good for particular patients. Second, belief in the sanctity of life or respect for life implies that denying or withdrawing care on the basis of cost debases the fundamental sanctity of life and violates the principle of respect for life which sees life as a moral object worthy of respect. Third, human life is priceless; it has a moral status which in principle precludes assigning it any exchange value. Hence, considerations of cost insofar as they assume that human life has exchange value must always be rejected on theoretical grounds.

I do not argue that these are the only possible objections, but rather simply assume that these three objections are sufficiently common to merit critical analysis. The goal of my analysis is to show that on theoretical grounds each of these objections fails to rule out economic considerations in medical practice. This failure is of interest because it demonstrates not only that consideration of cost in medical care is not ethically prohibited as such, but also that it is in some circumstances ethically required. Thus, I will argue that problems such as cost containment and rationing of medical care are intrinsic problems for medical ethics rather than dilemmas foisted on medicine by recent economic and political circumstances. The failure of medical ethics to address adequately these issues is due at least partly to the pervasive influence of the three common objections. These objections must be definitively set aside if medical ethics is to contribute to public discussion of the issues associated with the economics of medical care. The intent of this essay, then, is modest: to reject certain standard objec-

G. J. Agich and C. E. Begley (eds.), The Price of Health, 23–42.

tions. In Sections I—III I consider each objection in turn and conclude in Section IV with a brief discussion of how and why economic concerns can be incorporated in medical ethics.

I. TRADITIONAL MEDICAL ETHICS

Traditional medical ethics views the individual physician-patient relationship as structured by a unique principle of beneficence, namely to provide the utmost benefit to a particular patient. This view holds that the principle of beneficence obligates the physician to do all that is possible for patients irrespective of cost.[1] Such a belief repudiates the view that modern medical practice requires that physicians make hard choices and face squarely the fact that not all can be done because the costs may be too great.[2] On the traditional view, it is simply anathema that a physician should recommend or carry out a course of action that is anything short of "the best".

For present purposes I ignore the fact that this view may be partly motivated by physicians' psychological inability to admit defeat or even to provide palliative, preventive, or rehabilitative care in lieu of attempting outright and heroic cures. I also ignore the fact that institutional arrangements and economic incentives associated with the widespread availability of third-party medical insurance (whether private or public) have insulated health care providers and patients alike from the economic consequences of medical care decisions. Although these facts actually may go a long way toward accounting for the resilience of the "traditional" view of medical ethics, I am less interested in the motivations for holding the view than in exploring its implications for the inclusion of economic considerations into medical practice. For convenience I have developed my argument as a criticism of the traditional position as defended by Edmund Pellegrino and David Thomasma in their book, *A Philosophical Basis of Medical Practice* [17].

On their view, medicine is traditionally a moral enterprise which normatively conflicts with economic considerations. Medical economics, they point out, is rooted in three classic observations; first, resources are scarce relative to wants; second, resources have alternative uses; and third, people have different wants and assign different values to them ([17], p. 267). The essential problem for economics, in their view, is to allocate resources in an efficient manner so as to meet human needs: "If needs judged as goods are values, as we have suggested, then medical economics must deal with health as a value" ([17], p. 267). The theoretical conflict which they see between medical ethics and economics, then, is that medical ethics requires commitment not only to the best interest of the patient – which, of course, in some circumstances could require abandonment of medical care in order to pursue other goods – but an absolute commitment to the value of health for the particular patient. Medical economics, on the other hand, is concerned with the efficient use of medical resources in the aggregate. As such, it substitutes a decision-making procedure that abstracts from the particularity of the individual case in order to structure a practice that best meets certain rational decision-making criteria or is designed to maximize utility. This process may serve the good of patients generally or of society, but not primarily the good of the individual patient.[3]

Pellegrino and Thomasma seem to view the conflict between economics and ethics in terms of a society/individual distinction. What is problematic about economics on their view is that it tends to substitute concern for society or patients generally for concern for the particular patient. Even though this view of economics is overly simply, I am more concerned with the positive moral argument which they must advance in order to exclude economics from medical care. This argument has at least three stages which parallel their tripartite theory of medical morality: the fact of illness, the act of profession, and the act of medicine ([17], pp. 207–212). In the first stage, it must be established that the pursuit of health is morally required. It is not enough simply to establish that health care is morally permissible, because these authors assert that health is an absolute value which structures the moral obligations of the physicians. At least for physicians, health is an absolute moral good ([17], p. 187–89). In the second stage, it must establish that it is morally permissible to employ those resources necessary to pursue health in the care of individual patients. That is to say, the pursuit of health absolutely should not conflict with other moral obligations or rights. In the final stage, it must be established that the moral authority of the physicians to pursue treatment that is good and right for particular patients logically precludes consideration of cost.

Pellegrino and Thomasma themselves are clear that their account of medical morality stems from viewing health as a value which they base on the ontological conditions of the living body and the so-called fact of illness ([17], p. 187). Principles of medical ethics are thus descriptions or elaborations of a practice based on the fact of illness and the value of health. In these terms, they view illness as an ontological assault which impairs the body-self unity in a unique way. But this reading actually gives medicine only a *prima facie* normative foundation, because the concept of illness is itself dependent on health as a value. What is different about the ontological assault of illness on the body-self unity is the way in which illness removes the primacy of freedom to deal with all other life situations. This point is expressed in the adage: "If you have your health, then you have everything". Further, in illness, restoration of freedom is not a straightforward matter, according to them, because "essential existential mechanisms for coping with all other exigencies have been compromised, and more essential than that, we face the threat of loss of life itself, or we are suddenly asked to live a life not worth living" ([17], p. 209). Because of this special dimension of anguish, Pellegrino and Thomasma argue that healing cannot be ethically classified as a commodity or as a service on a par with other economic commodities and services.

Pellegrino and Thomasma thus distinguish illness as an ontological assault from the loss of other freedoms and argue for the ontologically unique way that illness compromises freedom ([17], pp. 187 and 207–208). This argument, however, is really an explication of illness from the *point of view* of medicine — from an ideology of medicine — rather than an argument that successfully establishes loss of health as more fundamental morally than loss of any other freedom. Pellegrino and Thomasma, in fact, do not provide an independent argument for the ethical priority of illness over other losses of freedom.

Indeed, it is hard to imagine how such an argument might be framed. It certainly would have to deal with the following example. A political prisoner or victim of terrorists is subjected to a situation which threatens life and essential existential coping mechanisms. Release of the prisoner or captive will not necessarily simply remedy the existential harm done. Now Pellegrino and Thomasma could argue that this example points to the fact that the harm done by virtue of imprisonment constitutes an illness rather than another kind of deprivation of freedom insofar as it impairs essential existential coping mechanisms and as such requires remedial medical or psychiatric treatment; hence, illness would still be a unique kind of ontological assault. Such a response, however, would beg the question. One could always *say* that the example constitutes illness just insofar as the particular example is telling. For the Pellegrino and Thomasma position to be taken seriously, they must either mean that illness *causes* a unique ontological deficiency or *is* a unique ontological deficiency. Their account is, however, not causal, but essential; they mean that illness *is* a unique ontological deficiency.

Imprisonment by terrorists is, therefore, a useful counter-example, because it can certainly place a prisoner in a situation in which essential coping mechanisms are compromised and in which the prisoner is threatened with loss of life and is made to live — possibly for long periods of time — a life not worth living. The prisoner's body is placed in a circumstance that frustrates his freedom. It no longer carries out his actions in the world. It is coercively made to serve the terrorists' propaganda purposes rather than the prisoner's own will. Suffering under such conditions is certainly morally compelling, but is not for that reason *illness,* nor is it for that reason a proper concern of professional medical morality. The example makes it difficult to determine on what grounds the specific anguish in illness is claimed to be relevantly different from anguish associated with other losses of freedom in appropriate extreme cases.

But perhaps more importantly, the majority of illnesses hardly seem to fit Pellegrino and Thomasma's definition of an "ontological assault" for the simple reason that they are self-limiting and constitute minor, transient irritations, whereas other serious losses of freedom (which would hardly be recognized as illnesses or as proper objects of medical practice) do constitute relatively more serious assaults on freedom. Pellegrino and Thomasma's argument therefore fails insofar as it is dependent on a description of illness as a *unique* ontological assault. As a result of the failure of this argument, the fact of illness cannot serve as the basis for precluding economic considerations in the physician-patient relationship.

Pellegrino and Thomasma's argument against regarding medical care as a commodity, however, also relies on the contractual or fiduciary character of the physician-patient relationship, which they term the act of profession to do what is right and good for the individual patient. Their argument is ethically compelling only if two conditions can be met: first, the promise of absolute care must be mutually acknowledged or accepted by patients. As Pellegrino and Thomasma themselves point out, the promise to heal the patient acquires "a normative force only when health and healing are ranked as primary by *both* [italics added] physicians and patients" ([17], p. 187). In short, implicit in the traditional model is agreement or consensus that health and health care are relatively more important than non-medical goods and values. Along with this assumption, individual patient informed consent always is presumed to be present or at least to be non-problematic.

Even when this first condition is met, a second condition must be met — namely that resources utilized can be justly or morally claimed by individual physicians and patients. There are at least two aspects to this second condition which have been captured by Charles Fried when he observes: "Surely it would violate both negative rights and the notion of personal liberty in the bestowal of friendship to forbid a doctor from doing good to whomever he chose so long as he used his own time and resources" ([11], p. 180). Now if that were the extent of the Pellegrino and Thomasma claim, I would readily grant it. But as Fried goes on to note: "Now, there is a large problem about the resources (medicines, hospital beds, and so on), but the point is that all of the doctor's *time* is this own, because *everybody's* is his own" ([11], p. 180). Here, Fried wants to establish the *freedom* of physicians to act charitably, whereas I am primarily concerned with the question of the moral entitlement to the resources used. Thus, the important point to note is that physicians' time is not as significant a cost component in medical care as is their role as gatekeepers of health care resources.

As gatekeepers they control access to myriad medical resources which are not their own. For this reason, physicians are free to bestow care on their patients only insofar as the care is theirs to give justly. I argue later in Section IV that the fact that health and health care are pursued by moral agents does give health and health care a moral significance; nonetheless, this fact does not necessarily mean that devotees of medical care have a justifiable moral claim on resources which are not their own. For this reason, consideration of the cost of medical care is not simply a matter of prudence for physicians, but a moral responsibility if they are to respect the rights of others to resources that *are* morally their own. In addition, this point involves the recognition of the fact that questions of access, availability, and distribution of medical services are important ethical problems from which the traditional model abstracts in its exclusive focus on the immediate physician-patient relationship. To be sure, these issues are not simply matters of monetary costs, but they do involve considering costs and benefits in the provision of medical care, an option excluded by the traditional account.

In the final stage of the argument, it would have to be shown that physicians have the moral authority to ignore consideration of cost. In a sense, the authority of the physician on the traditional account depends on the concept of a good and right healing action which involves two components: first, technical and scientific knowledge and, second, value judgment about what is good and right for particular patients. Any therapeutic application of medical knowledge to a particular patient is always uncertain because of the "experimental" nature of clinical practice as Samuel Gorowitz and Alasdair MacIntyre have noted ([12], p. 265). For this reason, uncertainty is an important feature of medical practice. If there are alternative ways to pursue health, because medical knowledge is always empirical and therefore uncertain, then value choice must enter. In addition, decision making under conditions of uncertainly requires at least a judgment, if not a calculus, of how risk aversive individuals are and what goals they want to pursue. The question is whether patients should yield to physician authority to decide such matters as the traditional model seems to require.

Robert Veatch has observed that a coincidence of patient values and physician expertise may actually have occurred in recent history during the 1940s and 1950s ([23], pp. 129—130). During that time the paradigm illnesses were infectious diseases. Patients needed experts who knew about the medical facts. If the patient had pneumonia and a drug existed such as penicillin to cure the patient's fatal illness, what was needed was precisely medical expertise, a good and right healing action, an action that was, relatively speaking, low cost yet highly effective. This action was justified because there was a consensus on beliefs and values regarding care; patients wanted to be treated and needed physician expertise to achieve their goal. Now in the 1980s, however, cancer, heart disease, and other chronic diseases are dominant, and there is no consensus regarding which available medical interventions are best. Uncertainty, which is always present in medicine anyway, becomes poignant. The choices of surgery, radiation, and exposure of patients to toxic chemicals in treatment of cancer are not predicated on a consensus that these treatments are worthwhile. Similarly, for heart disease, physicians can operate, transplant, medicate, and try to change patient lifestyles, but the question is whether these interventions are worth-while. To be sure, that is partly to be settled by appeal to scientific and technical evidence, but it also involves fundamental beliefs about and attitudes toward health as a value. The central question is whether physicians should have the final authority in deciding such matters simply because of their technical expertise. As Veatch expresses it:

> I concede that health is a legitimate value. I even concede that medical professionals should be the priests of that value. I submit, however, that no one in his right mind would conclude that those who are custodians of a particular value should bear the authority for resolving disputes over the relation of that value with other values leading to one's integrated wholeness ([23], p. 133).

It should be clear, then, that at least when significant uncertainty exists regarding what is the medically proper or best course of treatment, it is important for patients to consider whether any particular treatment option is worthwhile for themselves in terms of their own values and preferences; a component of such decision making is to take account of alternative uses for the resources. This conclusion is in conflict with traditional medical morality, which requires an uncompromising commitment to the pursuit of health for each patient.

Economic considerations thus are not ruled out by the traditional view of medical ethics — at least as that view is defended by Pellegrino and Thomasma. The problem with the traditional account is that by positing health as normative for the physician-patient relationship, it fails to acknowledge the existence of competing values that often figure in medical decision making, especially when pursuit of health itself becomes a disvalue for the individual or for society, and it tends to isolate decisions regarding individual patients from consideration of the consequences of those decisions on others.

II. SANCTITY OF LIFE AND RESPECT FOR LIFE

There is a second source for the view that medicine precludes consideration of economic cost. This view has developed in a long and pervasive tradition within the medical profession and Western culture generally. It finds expression in various non-equivalent phrases such as "respect for life", "sanctity of life", "right to life", and so on ([7], [8], [16], [21], [22], [26]). This view holds that denying or withdrawing care, even marginally beneficial care, on the basis of cost-worthiness criteria debases the fundamental sanctity of life and violates the principle of respect for life which sees life as a moral object worthy of respect.

Given the diversity of expressions of the central idea of this tradition, it is important to note that the sanctity/respect for life tradition can be seen as precluding consideration of economic cost in medicine only if it can be interpreted as providing sound normative or ethical judgments about life. In other words, over and above articulating a feeling of respect for life or an attitude of awe at the sanctity of life, this tradition must contain arguments by which the objections to economic considerations are developed and defended. I do not find, however, that the required arguments are present. As Owsei Temkin has noted, there are even basic problems in interpreting the significance, much less the meaning, of this view for medicine:

> Medicine does not deal with life per se; this is the biologist's concern. To physicians, "respect for life" has meant "respect for human life". Medicine has not stood for vegetarianism, and animals have been sacrificed in experiments aimed at saving or improving human life. Respect for life in medicine has, moreover, revealed itself as a complex and paradoxical idea, not free from contradictions ([22], p. 15).

William Frankena has argued that medicine is committed to a *non-comprehensive* respect for life, particularly a respect for the individual human (bodily) life, but he also acknowledges that even with this restriction it is not clear what "respect" means or what "individual human life" means ([10], pp. 25–26). Consider the following possible meanings of "individual human life". The phrase could refer to the *quantity* of life, either the length of an individual's life or the number of individual lives in existence, or to the *quality* of life. The quantitative view assumes that the value of life is capable of measurement. Such a view is possibly compatible with utilitarian approaches. On such an interpretation, "respect for life" seems to support rather than reject those consequentialist approaches which turn on quantitative measurement of the qualities of human life and which involve assessments of cost. The latter quality of life view, however, is not strictly speaking concerned with life at all, but rather with the quality or good of individual human life, for example, freedom or happiness. In this case, it is not life as such that is being valued, but rather whatever *in* life constitutes the means to achieve the good *of* life. This latter interpretation of respect for life, however, conflicts with the main thrust of the sanctity of life position which prohibits valuing life instrumentally in terms of the qualities it possesses or exhibits ([16], pp. 21—22). Thus, the sanctity of life position involves a general endorsement of the view that life itself is morally valuable without clearly defining what "life" means.

Presumably, an increase in the quantity of life would be regarded as a good thing by proponents of this view. What is not clear, however, is whether this view *requires* such an increase. At bottom, the sanctity of life position seems to presuppose rather than justify any explicit meaning of the term life. In fact, the view seems mainly to depend on a variety of religious attitudes towards life. Such attitudes of love towards life and of living, however, do not themselves constitute *respect* for life in any moral sense. As Frankena notes, respect for life which is predicated on the fact of love of life and of living does not include any normative or moral value judgment about its object as such ([10], p. 28). Instead, it simply expresses an affection or a desire for life.

In the next section I argue that it is incorrect and misleading to speak of "respect" for human life unless one really means respect for human *personal* life, because it is personhood and not human biological life which is the proper object of our moral regard. It is not at all clear, however, that such an interpretation is central to the broad tradition of respect for life or sanctity of life. Unless it can be shown that phrases like "sanctity of life" or "respect for life" involve a normative or ethical value judgment about life rather than simply expressing a desire or love for life, it is difficult to interpret the attitudes implied in the phrases as conveying normative ethical force. If this tradition lacks normative ethical force, then it surely cannot be taken as an adequate, much less a decisive, moral objection to the consideration of economic cost in medicine.

In this regard, Frankena has argued that two rather different answers to the question of the normative status of "sanctity of life" and "respect for life" have to be distinguished:

> One is to say that human life is "sacred", "sacrosanct", or even "holy"; the other is to say that live human beings have certain rights, that it is wrong to treat them in certain ways. These two lines of thought are often confused or conflated, but, even if the first entails the second, they should be distinguished ([10], p. 30).

In other words, the sanctity of life view involves two different approaches to the question of the normative foundation of the actions which it entails: a religious or protoreligious form, in which it is natural to speak of respect for human life as including a sense of awe or wonder that is akin to the awe or wonder which is reserved for the sacred or religious; and second, a more purely ethical form, in which it is not natural or at least not necessary to speak of such a feeling of awe.

On the first reading, sanctity of life unfortunately does not provide a firm normative foundation from which to reject consideration of cost. It is hard to draw sound implications for ethics from the feeling of awe alone. As K. Danner Clouser has observed, "neither command nor obligation follows from the fact that we feel a certain way about life" ([8], p. 27). Hence, it is hard to see why considerations of cost would be precluded by this view.

In this regard it is important to note that the second form of the doctrine of respect for life, namely that live human beings have certain rights, is or at least can be held independently of religion in anything like its traditional forms, and so this second approach avoids problems associated with deriving normative moral conclusions from expressions of religious belief or feeling. More importantly, however, because this approach is expressed in terms of rights, it requires that the rights be specified and justified which live human beings are said to have. This process of specification and justification, however, involves considerations which show that the doctrine of respect for life is either not up to the task of excluding cost considerations or logically involves reference to cost just insofar as it is expressed in terms of rights. For example, consider whether the doctrine of respect for life implies positive or negative rights.

One positive right which a doctrine of the respect for life might imply would be a positive welfare right to at least a basic or decent minimum of health care. If there were such a right, the doctrine of respect for life should figure centrally in its justification; but, so far, this task has proved not only to be formidable [2], but hardly seems helped by reference to respect for life considerations. Furthermore, positive rights always involve reference to cost, because of the scarcity of resources and the fact that alternative uses of resources might also be matters of rights. (A right to health care, for example, can lead to conflicts with property rights.)

It might be argued, however, that respect for life only implies negative rights, such as a right to be left alone, a right to non-interference. Others would then have a *prima facie* duty not to intrude on the individual's solitude, not to touch the individual, and so on. Here, respect for life would simply require forebearance from intruding into the life of individuals. As Charles Fried has pointed out, the intuitive notion is that negative rights do not confront the problem of scarcity, do not involve competing claims to limited resources, because they require only forebearance from certain actions ([11], p. 112). However, this intuitive characterization of negative rights overlooks opportunity costs.

By asserting a negative right as a constraint on action, it becomes more expensive to pursue or provide other goods, some of which may be validly claimed as a matter of positive right. Negative rights, then, also have costs and these costs are ethically relevant. They can make pursuing good purposes more difficult or even impossible if certain means are prohibited. It may be impossible, for example, to conduct a clinical drug trial without deceiving the subjects about the risks and purposes of the treatment they are receiving ([11] p. 113). Thus, a negative right not to be lied to might make clinical investigation costlier insofar as it precludes deception in research. This point stands even if society firmly believes that the costs associated with requiring and maintaining informed consent are worth bearing.

Negative rights involve costs not only when compared with positive rights, but also when compared with one another insofar as in the real world negative rights require a social system of enforcement, a system that will protect negative rights. To secure my right to be left alone requires some social or institutional mechanism to assure that my solitude will not be intruded upon. The establishment and maintenance of such social or institutional mechanisms to protect negative rights involve real costs that must be accounted for. Thus, negative rights, like positive rights, can hardly be said to preclude consideration of cost.

The tradition of "respect for life" or "sanctity of life", then, on any of the interpretations just considered affords a rather insecure foundation for objections to considering economic cost in medical care. These interpretations therefore are compatible with the view that at least under some circumstances cost considerations are not ethically proscribed. Even if my interpretations of this tradition are wrong, it would still not be clear that the traditional use of "sanctity" or "respect" for life provides the requisite normative force to prohibit attention to cost. Until compelling argument is advanced, it seems safe to conclude that the sanctity of life tradition does not provide a formidable objection to considering economic cost in medicine.

III. THE PRICELESSNESS OF LIFE VIEW

A third and weighty objection to the introduction of economic considerations such as cost-benefit or cost-effectiveness analysis into medicine is that life, particularly human rational life, is priceless. The *locus classicus* for this view is Immanuel Kant's claim that human beings have a dignity but not a price ([15], p. 53; Ak. 434). On this view, human beings cannot be evaluated in monetary terms; they are literally priceless. Two preliminary points need to be stressed with respect to Kant's view: first, Kant speaks of human beings rather than human life, a point missed by some utilitarian critics who view human life in quantitative terms and speak of "a willingness to exchange the length of one's life for its quality in the sense of fulfillment of desires".[4] By the term *human being,* Kant means human *personal* life which is founded on autonomy and not life as a quantitatively extended "life span". As the context of Kant's distinction makes abundantly clear, human *personal* life has dignity because it alone is the autonomous subject of morality. Thus, human persons cannot have exchange value of any kind, but rather have a dignity owing to their unique status both as subjects of morality and as members of a community of such subjects.

Second, by the term *price* Kant means both that human beings lack monetary value and that individual human beings are not equivalent. As he puts it: "Whatever has a price can be replaced by something else as its equivalent" ([15], p. 53; Ak, 434). Human *personal* life, then, is irreplaceable just insofar as it is the autonomous subject of morality; as the bearer of certain qualities such as skills or talents, however, individuals are replaceable but only when viewed in terms of their objective (empirical) attributes, not in terms of their subjective (noumenal) status as autonomous moral subjects.[5]

The ethical issue, for example, in replacing a secretary does not concern the fact that human beings are being valued or priced for their typing or transcription skills, but rather whether the individuals are treated justly or whether their rights have been violated in the process of hiring and firing. In such a case, one replaces the lost skills or services, not the person as such. Therefore, it is incorrect to say that ethics precludes considerations of cost for Kant; rather, it transcends them. Kant's claim is simply that human beings have an intrinsic worth which cannot be ethically valued in economic terms and which precludes the assignment of any kind of exchange value on human *personal* life, but that does not preclude the assignment of value on attributes of human life such as skills ([15], pp. 53—54; Ak. 434—35). The point that should be stressed is that the pricing or valuing of features of human life and the practices within which such pricing occurs must not be conducted in such a way that morality, namely the personhood of individuals, is violated.

This argument only establishes that the pricing of the qualities of human life is morally permissible if the practice within which it occurs is morally acceptable and if the particular pricing of skills, for example, does not violate morality. That is, if hiring individuals to perform secretarial tasks is morally acceptable, then it is acceptable to value secretarial skills according to certain standards and to set a market price on performance which meets these standards. In other words, more highly skilled candidates could be awarded a higher rate of pay than candidates with a lower level of skill. This would not mean, however, that it is morally acceptable to pay more for a skilled slave than an unskilled slave since slavery is itself immoral. The significant difference consists in the fact that the secretary receives reimbursement for his work, whereas the slave does not. The secretary is free to sell or withhold his secretarial skills from the market, whereas the slave is not. Slavery is immoral, then, not because human life quantitatively regarded is priceless, but because the slave is denied the status of personhood, namely autonomy. Even within a morally acceptable practice such as the employment of free individuals for a wage, there could be morally objectionable actions such as denying a pay raise for an employee's refusal to supply sexual favors or awarding a large raise for supplying sexual favors.

The case is similar for health care. If needed but costly care were denied or withdrawn solely on the basis that the individual person in question was not worthy of it, say because the individual was poor or suffered from a disease such as AIDS, which unfortunately carries with it strong social stigmatization, morality would be violated. However, a judgment that the cost of a heart transplantation is not worthwhile for an individual who stands to benefit from it only marginally or not worthwhile because of a fairly-determined public policy to fund alternative interventions does not necessarily violate morality, even though consideration of economic cost figures prominently in the justification of these actions. Of course, this does not sanction abandonment of patients, because compassion would still require that minimally decent care such as the use of palliation and psychosocial support for the patient and family be provided.

Kant's view that human *personal* life is priceless thus does not entail the claim that aspects of human life cannot be valuated in economic terms. Rather, it entails the view that economic valuations of life (or the qualities of life) are morally acceptable only insofar as such valuations are either permitted by morality or are required by morality. Moral considerations serve to constrain considerations of cost. Kant's view of the priceless character of human life, therefore, does not rule out considerations of cost in medicine and so cannot be taken as a substantive objection to the introduction of economic analysis into health care.

IV. CONCLUSION

If the arguments just advanced are sound, then neither traditional medical morality, the sanctity/respect for life tradition, nor the view that human life is priceless can be seen as ruling out considering cost in medicine. What positive conclusions can be drawn from this negative line of analysis?

The view I want to defend is derivative from Kant's. I acknowledge the priceless character of human *personal* life as a basic axiom of moral theory, yet recognize that in many circumstances this does not preclude the pricing or differential valuing of various attributes or qualities of human life. One of these aspects is health. Health, and those things necessary for its attainment or preservation, are instrumental goods, goods worth pursuing to attain human welfare or happiness. As such, they do not occupy a special place in morality except insofar as they are freely pursued by persons. Because it has instrumental value, health is capable of being priced. In that regard it is strikingly different from human *personal* life as such. Morality does not require that one be healthy or engage in health promoting activities or pursue needed medical care; that is more a matter of prudence than morality. *Prudence* is skill in the choice of means to attain one's happiness or welfare. According to Kant, "the imperative which refers to the choice of means to one's own happiness, i.e., the precept of prudence, is . . . only hypothetical; the action is not absolutely commanded but commanded only as a means to another end" ([15], p. 33; Ak. 416). This is distinguished from the Categorical Imperative which commands absolutely and which forms the basis for the Kantian moral doctrine of "respect for persons".

As a matter of fact, people do price health and those things required for health in the daily course of their lives. To do so is perfectly consistent with their status as moral subjects. To be a moral subject implies the capacity to choose freely and rationally. Such a capacity itself implies a system of wants, and so the concept of a moral person itself implies a desire to choose and attain a particular conception of the good. Happiness and human welfare have moral worth, then, because they are freely pursued by moral beings. Therefore, respect for persons entails valuing the goods which moral beings have chosen just because they have chosen them and attributing moral worth to human happiness or welfare insofar as they are the goals of moral beings. As Charles Fried has put it:

> Moral personality consists . . . of the capacity to choose freely and rationally. But the capacity to choose implies a system of wants or ends, and so naturally moral personality implies a desire for happiness, a desire to choose and attain a conception of the good. So on the one hand, happiness has value and we are morally required to value the happiness of our fellow men. On the other hand, human happiness has moral worth only because it is the happiness of moral beings, and therefore happiness has moral worth only as it is pursued within the constraints of the moral law ([11], p. 123).

Free choice also has costs. It entails both that alternative goals will not be pursued and that effort or resources will be expended in the pursuit of the chosen goals. When the resources employed are justly one's own and one is not morally bound to employ them for other purposes, for example, not bound by prior promise, contract, or responsibility to use them for the benefit of others, then respect for persons requires acknowledgment of the goals freely chosen by moral subjects. Respect for persons does not require, however, that health, health care, or any other instrumental good be pursued as absolute values. Rather than precluding the consideration of cost, respect for persons provides both the context and sets constraints which determine the ethical status of cost. For example, it is widely recognized that refusal of life-sustaining treatment by dying patients should be honored ([20], p. 23; [18]). Such recognition is a tacit acknowledgment not only that the costs of treatment may outweigh its benefits or may impose undue burdens in certain circumstances, but also that cost as such is an ethically relevant consideration.

This conclusion, however, does not entail a free-market approach in which the price of health and health care is set solely as an exchange commodity. Health is a morally relevant good not simply because moral agents happen to desire or prefer health, but because moral principles such as beneficence, charity, and justice support obligations to aid the needy. Moral action, like all human action, involves effort. Economic costs are involved as soon as it is recognized that individual, non-market effort to care for a sick person can be generalized or universalized in terms of communal efforts dedicated to aiding the needy. The isolated individual Good Samaritan who cares for an individual in need gives rise socially to either formal or informal societies of "Good Samaritans" whose purpose is to benefit those in need of medical care. The history of medical practice points to the inextricable involvement of individuals with one another in communities of caregivers. Modern medicine is such a community of organized, institutionalized modes of intervention. Thus, economic costs, as exchange, market-determined costs, are relevant morally because individual persons belong to communities in which goods and services are created and exchanged freely. Economic considerations, then, are both related to, yet constrained by, the same features that give rise to morality, namely consideration of the larger community of moral subjects and the universe of moral value that structures that community.

Southern Illinois University,
School of Medicine
Springfield, Illinois

NOTES

[1] For a general discussion of the scope of beneficence in health care, see [1], pp. 183—198).

[2] I do not mean to imply that the view that medicine is constrained by "hard choices" is itself a particularly modern view. To be sure, the language of hard choice of competing technologies and the uncertain outcomes associated with their use is contemporary, but the problem of uncertainty and the recognition of the finite limitations of medical knowledge is classical. What seems particularly contemporary about the problem is the apparent reluctance of physicians to accept limitations inherent in medicine. This is expressed both in a slavishness to the technological imperative and to the belief that medicine has a positive duty to prolong life. This latter duty, as Darrel Amundsen has

shown, is itself without classical sources [3]. The duty that seems to have guided much of medicine until the modern era was a duty of nonmaleficence expressed in the ambiguous epithet *primum non nocere.* See [4]. In the modern era, nonmaleficence seems to have been superceded by a principle of beneficence, namely that it is always better to try to do good than to try to avoid harm or evil. Coupled with the faith that underlies the technological imperative, the principle of beneficence unfortunately seems to give rise to a hope in satisfactory outcomes without the corresponding sense of responsibility to measure whether the outcomes are indeed better or worse.

[3] Pellegrino and Thomasma argue that only when the patient is a danger to society, such as a patient with a contagious disease or a dangerous psychotic, does the physician have the responsibility to carry out social policy ([17], p. 270). But it is not clear on what ethical principles they justify the physician's responsibility in the case of contagious disease or dangerous psychosis. In fact, Pellegrino and Thomasma gloss over this point by indicating that the "good of the patient" may itself dictate the same action which is good for society as well ([17], p. 270). Indeed, it may. But this response ignores cases of real conflict. Hence, it is not a sufficient justification for an exception to their medical ethical principle of primarily benefitting a particular patient. They seem, however, to be unaware of this fact. One possible line of defense that seems closed to them is the argument that individual patient welfare is inextricably bound up with social welfare. In such an argument, social welfare logically would have to be given coordinate consideration with the welfare or good of particular patients. They seem to want precisely to block coordinating social welfare with the welfare of individual patients on the moral ground of the *uniqueness* of illness as an ontological assault. Indeed, they go so far as to argue that the physician "also has the responsibility to resist those policies when they are detrimental to the welfare of his patient" ([17], p. 270).

An important question which has to be considered here is whether Pellegrino and Thomasma think that individuals who *freely* assume risks to health can make moral claims to health care as a matter of rights. Certainly, physicians (and others) should be free to treat (or support treatment for) such individuals out of charity, but this fact alone does not justify the expenditure of society's resources ([24], pp. 50—55; [25]), nor, as I argue below, does it exempt physicians from social welfare responsibilities. I do not want to foreclose the possibility of such a justification, but only point out that the case has hardly been made by the "traditional" account. Recent work by Allen Buchanan [6], Norman Daniels [9], and others (see [5]; [18]) illustrates the richness of and difficulties associated with attempts to justify a right to at least a basic minimum of health care. The traditional account assumes such a justification, but does not provide it.

[4] (See [4], p. 26). To speak, as does Michael Bayles, of a willingness to exchange the length of one's life for its quality in the sense of fulfillment of desires is troublesome also because not all desires are on a par. On Kantian grounds one would have to distinguish those "desires" that generate a maxim which is universalizable from those desires which are incapable of such universalizability because they are merely subjective or "pathological" as Kant says. Kant's claim that human *personal* life is priceless indicates that morality sets constraints upon the pricing or exchange of the length of one's life for the fulfillment of choice or desire. Only desires capable of universalization permit exchanging the length of one's life for fulfillment of the desire; otherwise, one's life would be subject to exchange for *any* desire no matter how aberrant so long as the desire were sufficiently strong. The latter alternative, Kant argues, violates morality.

[5] Kant does not mean that individual persons are irreplaceable in terms of the qualities or features of their lives. Human *personal* life is irreplaceable in the sense that no specific empirical qualities of life can "stand in for" or replace autonomous moral subjects. This point is important because Kant cannot be interpreted to mean that human personal life is irreplaceable in any terms, because morality precisely requires the universalizability of the maxim of action. In a sense, it could be argued that the individual moral agent is "replaced" by other moral agents under the universalization procedure. Kant's view that human *personal* life is irreplaceable must be taken to mean that it is not capable of being replaced by empirical qualities such as skills or talents. Hence, Bayles ([4], p. 24) is correct when he concludes — against Kenneth Henley [13] — that the concepts of irreplaceability and pricelessness are not the same. The point Bayles seems to miss — or simply rejects without argument as "metaphysical dualism" ([4], p. 24) — is that Kant's view of irreplaceability depends on the fundamental distinction between a human being qua moral subject (person) and an empirical living human being. Without a distinction such as this, the move to a utilitarian calculus of costs and benefits is made easier, though it is attended with other serious difficulties, as other papers in this volume make clear.

BIBLIOGRAPHY

[1] Abrams, N.: 1982, 'Scope of Beneficence in Health Care', in E. E. Shelp (ed.), *Beneficence and Health Care,* D. Reidel Publishing Co., Dordrecht, Holland, pp. 183—98.

[2] Agich, G. J. and Begley, C. E.: 1985, 'Some Problems with Pro-Competition Reforms', *Social Science and Medicine 21,* 623—630.

[3] Amundsen, D.: 1978, 'The Physician's Obligation to Prolong Life: A Medical Duty Without Classical Roots', *Hastings Center Report 8,* 23—30.

[4] Bayles, M. D.: 1978, 'The Price of Life', *Ethics 89,* 20—34.

[5] Bondeson, W. *et al.*: *Rights to Health Care,* D. Reidel Publishing Co., Dordrecht, Holland, *in press.*

[6] Buchanan, A. E.: 1984, 'The Right to a Decent Minimum of Health Care', *Philosophy & Public Affairs 13,* 55—78.

[7] Callahan, D.: 1968, 'The Sanctity of Life,' in D. R. Cutler (ed.), *Updating Life and Death,* Beacon Press, Boston.

[8] Clouser, K. D.: 1973, 'The Sanctity of Life: An Analysis of a Concept', *Annals of Internal Medicine 78,* 119—125.

[9] Daniels, N.: 1981, 'Health-Care Needs and Distributive Justice', *Philosophy & Public Affairs 10,* 146—179.

[10] Frankena, W. K.: 1977, 'The Ethics of Respect for Life', in O. Temkin *et al., Respect for Life,* The Johns Hopkins University Press, Baltimore, pp. 24—62.

[11] Fried, C.: 1978, *Right and Wrong,* Harvard University Press, Cambridge.

[12] Gorowitz, S. and MacIntyre, A.: 1976, 'Toward a Theory of Medical Fallibility', in H. T. Engelhardt, Jr. and D. Callahan (ed.), *Science, Ethics and Medicine,* Institute of Society, Ethics and the Life Science, Hastings-on-Hudson, New York, pp. 248—74.

[13] Henley, K.: 1977, 'The Value of 'Individuals', *Philosophy and Phenomenological Research 37,* 345—52.
[14] Jonsen, A.: 1978, 'Do No Harm,' *Annals of Internal Medicine 80,* 827—832.
[15] Kant, I.: 1959, *Foundations of the Metaphysics of Morals,* L. W. Beck (tr.), The Bobbs-Merrill Co., Inc., Indianapolis and New York.
[16] Keyserlingk, E. W.: 1979, *Sanctity of Life or Quality of Life: A Study Written for the Law Reform Commission of Canada,* Minister of Supply and Services, Ottawa, Canada.
[17] Pellegrino, E. and Thomasma, D.: 1981, *A Philosophical Basis of Medical Practice,* Oxford University Press, New York and Oxford.
[18] President's Commission for the Study of Ethical Problems in Medicine and Biomedical and Behavioral Research: 1983, *Deciding to Forego Life-Sustaining Treatment,* U.S. Government Printing Office, Washington, D.C.
[19] President's Commission for the Study of Ethical Problems in Medicine and Biomedical and Behavioral Research: 1983, *Securing Access to Health Care,* Appendix II, U.S. Government Printing Office, Washington, D.C.
[20] Robertson, J. A.: 1983, *The Rights of the Critically Ill,* Bantam Books, New York.
[21] St. John-Stevas, N.: 1963, *The Right to Life,* Holt, Rinehart and Winston, New York.
[22] Temkin, O.: 1977, 'The Idea of Respect for Life in the History of Medicine', in O. Temkin *et al., Respect for Life,* The Johns Hopkins University Press, Baltimore, pp. 1—23.
[23] Veatch, R. M.: 1982, 'Medical Authority and Professional Medical Authority: The Nature of Authority in Medicine for Decisions by Lay Persons and Professionals', in G. J. Agich (ed.), *Responsibility in Health Care,* D. Reidel Publishing Co., Dordrecht, Holland, pp. 127—37.
[24] Veatch, R. M.: 1980, 'Voluntary Risks to Health: The Ethical Issues', *Journal of the American Medical Association 243,* 50—55.
[25] Wikler, D. I.: 1978, 'Persuasion and Coercion for Health: Ethical Issues in Government Efforts to Change Life-Styles', *Milbank Memorial Fund Quarterly/Health and Society 56,* 303—338.
[26] Williams, G.: 1958, *The Sanctity of Life and the Criminal Law,* Faber and Faber Ltd., London.

PART II

COSTS AND BENEFITS IN MEDICINE: SOME PHILOSOPHICAL VIEWS

E. HAAVI MORREIM

COMPUTING THE QUALITY OF LIFE*

I. INTRODUCTION

The title of this paper invites one to raise an eyebrow. Quality of life is a notion which we want to measure with precision, but seemingly cannot. We want to measure it for a variety of reasons. As individuals we wish not just to live long, but to live well. In seeking health care, for example, we wish not simply to survive, but to ease our pain, to ameliorate our handicaps, to shorten the course of our illnesses, to return as quickly as possible to the pursuit of our life plans and valued activities. As a society we similarly formulate some conception of our collective well-being, which we then pursue through public policies promoting, for example, health, education, defense and national culture. In all these activities, we must rely on judgments about the quality of life. These judgments may be rough, and they may not even be explicit; yet, as individuals or collectively as a society we cannot appraise our current lives or make decisions about our future without at least determining which states of affairs are acceptable and which not, and which alternatives are preferable and worthy of pursuit, which to be avoided. Without such basic judgments of value and priority, we would drift aimlessly, making decisions arbitrarily and ineffectually, wasting effort and resources.

In this sense we routinely can and do appraise and compare qualities of life. At the same time, though, it seems both impossible and undesirable to try to measure quality of life with precision. How could it be possible to place a numeric value upon the having — or the loss — of hearing or upon the specific pleasure derived from listening to a symphony orchestra? Or, how could the value of saving a limb be tallied against that of teaching a retarded child to read? Thus, there seems to be a kind of paradox in the idea of measuring quality of life: we seem unable to measure quality of life, yet routinely we do; we must, yet perhaps we really should not.

In this paper I explore just one dimension of this paradox: an ambiguity in the notion of "quality of life", which includes two very

G. J. Agich and C. E. Begley (eds.), The Price of Health, 45–69.

different senses of the term, an objective and a subjective quality of life. My goal is to clarify these senses in order to be better able to determine whether, when, and how to measure quality of life. My concern here is limited. I do not attempt generally to describe the ideal outcomes of health care policy, or to assess what role quality of life analysis should play in preparing or in assessing that policy. Rather, I try to identify some particular points at which quality of life questions arise and how these questions should be addressed in light of the conceptual distinction between objective and subjective quality of life.

II. OBJECTIVE QUALITY OF LIFE

Objective quality of life consists of two components: a set of material facts and a shared societal evaluation of those facts. On one level, the factual component is fairly straightforward. It consists of the observable characteristics which describe someone's life.[1] Each person has a certain shape and color of body, (normally) a conscious mind, a particular range of functions and abilities for the uses of his body and mind, an environment which includes family circumstances, community, personal dwelling, and so forth. The list could be endless, of course, and an enumeration of the relevant material facts in a particular case will depend on which aspects of life we wish to assess. Thus, an inquiry into the quality of someone's life as specifically affected by hemophilia would include some investigation of the physical and social discomfort caused by the disease, but probably not the political climate of his local community. Reciprocally, the attempt to make a single, overall judgment about the 'net' quality of someone's life — for example, with an inquiry as to whether life is still worth living for a patient with a painful terminal illness — would require a broader conception of what constitutes relevant information.

Not all material facts are as conceptually clean or as empirically obvious as a person's height and weight. Some material facts are describable only with concepts which are philosophically loaded. Calling drug addiction a disease, for example, involves a normative judgment that the addict's body chemistry and behavior are an undesirable deviation from some description of normality — a description which is itself normatively loaded. In addition, there are implicit metaphysical commitments with any definition of drug addiction that refers to a person's 'inability to control' his drug-seeking behavior — or to the inadequacy of his 'willpower'. Wherever factual descriptions thus contain some philosophical assumptions, material facts will require reflective judgments as well as empirical observations. Under normal circumstances, nevertheless, it will be possible to identify and distinguish these statements of what is the case from the second component of objective quality of life — the normative evaluation of these states of affairs, the judgments whether what is the case is good or bad, desirable or undesirable.

This normative component includes both those judgments which we share (a) as human beings generally and (b) as members of a particular society. As human beings we share a broad set of judgments about the conditions which are minimally essential to render any life worth living and whose absence is *ipso facto* a seriously undesirable deficit. Joel Feinberg, for example, speaks of welfare interests:

> In this category are the interests in the continuance for a foreseeable interval of one's life, and the interests in one's own physical health and vigor, the integrity and normal functioning of one's body, the absence of absorbing pain and suffering or grotesque disfigurement, minimal intellectual acuity, emotional stability, the absence of groundless anxieties and resentments, the capacity to engage normally in social intercourse and to enjoy and maintain friendships, at least minimal income and financial security, a tolerable social and physical environment, and a certain amount of freedom from interference and coercion ([17], p. 37).

Similarly, Philippa Foot [18], J. Kleinig [27], and Norman Daniels [9] speak of normality, including species-specific functioning and a minimum of basic goods. (See also [20], [23].) T. M. Scanlon, in like fashion, speaks of central interests with which virtually everyone must be concerned ([40], p. 661).

To be sure, there may be disagreement about the particular elements on such a list, yet the important point is that there are at least some circumstances of life which, judged in and of themselves, are undesirable for any human being and some which are desirable. These circumstances involve such basic factors as adequate shelter, major mental and physical abilities, and the like. Whoever suffers serious deficits with respect to these basics of human life thereby can be said to have a poor objective quality of life.

Within any given society there will also be further social judgments about what constitutes a good or poor quality of life. As Lord Devlin has noted in another context, a society must be based on a community of shared ideas about what counts as living well or living poorly, and about which values and lifestyles it will promote and which it will discourage ([10], p. 32). Arguably, such shared standards are part of the very concept of a society insofar as organized societies are distinguished from mere aggregates of individuals living in physical proximity.[2]

These shared values further shape the normative content of objective quality of life judgments and render them at least somewhat culturally relative. In a society in which intelligence and self-reliance are considered to be important ingredients in the good life, for example, retardation is likely to be seen as a more serious deficiency than it would be regarded, say, in a community which emphasizes familial bonds and mutual support for all members. Similarly, the aesthetic standards by which we judge a person to be beautiful or handsome may vary from one society to another.

It is unnecessary to secure unanimous agreement on the entire list of societal values by which quality of life is judged to be good or poor. We need only to understand that, wherever there is consensus about any such judgment, we will be entitled to draw conclusions about persons' objective quality of life. Thus, while we may or may not agree whether wealth always represents a good objective quality of life, we probably can agree that desperate poverty constitutes an objectively poor quality of life. The impoverished person may have good friends who boost his social quality of life, but his economic quality of life is nevertheless agreed to be an adversity.

Objective in this sense is not intended to imply that the goodness or adversity of someone's quality of life is some sort of value-fact independent of the (changeable) public opinion which supports it. Rather, objectivity here consists in two features: (1) the material circumstances which can be intersubjectively agreed to exist; and (2) the normative judgment which is so widely shared as to constitute a social fact, as it is built into the community's language and is accepted as appropriate by all who use terms such as benefit, adversity, poor, and so forth.

Further, this is not to say that one will always be able to distinguish clearly between the descriptive and normative components. The fact-value distinction can be a ragged one. Nevertheless, that problem need not be of great concern here. In ordinary cases the distinction between the facts of a case and the normative judgment we make about those facts will be clear enough. It is one thing to be missing a limb, and another to determine that this is undesirable and how seriously so. Even where these two dimensions cannot be separated clearly, there is little need for concern. The important point is that there is social agreement that (a) a particular state of affairs exists or a particular event has occurred (under some shared description), and that (b) normatively it is to be regarded in a certain way.

Finally, we should note that because such societal judgments reflect a rough consensus, they can only render equally rough quality of life ascriptions. Refined comparisons between closely similar situations are not available. Thus, while everyone can agree that diabetes and hemophilia are each undesirable and that either is worse than some minor scrape, our collective normative standards are ill-equipped to determine whether diabetes poses a worse or better quality of life than hemophilia. Society's broad, consensus-based quality of life judgments stand in significant contrast to the highly personal, more refined judgments found in *subjective quality of life*.

III. SUBJECTIVE QUALITY OF LIFE

A. Competent Persons

Subjective judgments of quality of life, like objective, begin with the material facts of personal circumstances. Arguably, that factual base may be broader than is evident in objective ascriptions of quality of life. If we suppose that it is possible for a person to experience psychological reactions whose appearance is not directly under one's control such as fear or sadness, but whose presence one may be unwilling or unable to disclose to others, then we may agree that there are psychological facts which are not intersubjectively observable and which therefore must be considered separately from the facts on which objective judgments of quality of life are based.

The most important and truly distinguishing feature of subjective quality of life, however, consists in the source of the normative judgments which are applied to the facts — however broadly or narrowly we identify them. In the case of the competent person, these judgments stem from his capacity as an autonomous being to formulate his own beliefs and values, to judge his life conditions, and to make his life decisions accordingly. As Gerald Dworkin has noted, even when an individual has little or no control over the material facts of his existence, nevertheless as an autonomous being he can still decide how to regard these circumstances and what he will do in order to live with them ([12], [13]). Thus, one can regard the need for food as a hedonistic opportunity to feast or as a spiritual challenge to fast. Or one's blindness may represent a crippling disability or an opportunity for personal growth. Perhaps most importantly for our purposes, the capacity to be autonomous is the capacity to adapt to adversity, to find meaning and satisfaction under conditions which, objectively, constitute a poor quality of life.

In the clinical setting, subjective quality of life assumes a special significance. One's basic health condition and medical options establish the material givens and thereby the limits within which choices can be made. Yet, the autonomous individual can still determine what personal meaning he will find in his circumstances and which of the available courses of action he will take. To a person who values highly the ability to communicate orally, for example, radiation may be preferable to laryngectomy in the treatment of laryngeal cancer, even though the latter option offers somewhat higher chances for long-term survival [29]. Or a person diagnosed with operable lung cancer may favor radiation over surgery, even where the latter provides a greater chance of long-term survival, in order to avoid the immediate risk of intraoperative death [28].

By no means do I wish to imply that it is simple or straightforward for autonomous persons to determine what their current subjective quality of life is, or what they would like it to be. Competing values may foster serious ambivalence, and the factual uncertainties which are so common in the health care setting may cause considerable consternation. The illness which requires the decision making may prompt a reconsideration of the very values which would ordinarily guide that decision. A close brush with death, for example, might result in an individual's valuing his family more highly or rearranging important life goals. Further, there are conditions such as depression or metabolic imbalance which can significantly impair the ability to reason competently.[3]

Neither do I wish to imply that, because autonomy and subjective quality of life are of central importance in determining patients' interests, patient interests alone should determine what is to be done. Interests of other people are surely relevant. The burdens placed upon families of gravely ill patients and competing needs of other patients in the face of resource scarcity are also relevant. Rather, I mean only to suggest that in the assessment of a competent patient's personal quality of life, his own views and preferences normally carry far more weight than society's rough consensus.

B. Incompetent Persons

Matters are considerably different when evaluating people whose autonomy is seriously impaired or absent. Although impairments of competence appear in many varieties and degrees, I shall focus upon the more extreme situations. By definition, these people are unable to make the value and belief choices which are so central to subjective quality of life judgments. Yet they still may have subjective experiences, indeed, perhaps rich sentient lives in some cases, and thus have at least a limited subjective quality of life under my original definition.

Unfortunately, these people may be unable to communicate the content of that mental life. The severely retarded person may have minimal or no language skills with which to conceptualize his experiences for himself or describe them for us, and the seriously demented patient may be unable to maintain a train of thought long enough to entertain and answer questions. Nor can we always rely upon such persons' physical (re)actions as a basis on which to make inferences about their mental lives. A person in persistent vegetative coma, for example, may have wake and sleep cycles and cry out in what might seem to be a reaction to some inner stimulus. Yet if all of the brain functions required for consciousness and thought are absent, there is more reason to regard these simply as random vegetative movements than as evidence of subjective experiences.[4] As a result of all these obstacles, we may be quite unable to describe the (subjective) material facts of these persons' existence.

More importantly, if we cannot describe those material facts, then neither can we produce subjective or even objective evaluations of them. These persons' own subjective judgments are unavailable for obvious reasons (if indeed they can even be said to entertain subjective judgments or a subjective point of view), and it is doubtful that any observer could offer an adequate "substitute judgment". Where one cannot describe an individual's mental life, one can hardly evaluate that life, let alone evaluate it as the person himself would.

This same dearth of material facts also renders ordinary objective judgments about quality of life difficult, perhaps even impossible. While we may be able to appeal to such values as the general desirability of being alive or to the general desirability of avoiding pain, we often have little else on which to base a broader judgment. We would have little warrant, for example, to apply such sophisticated concepts as contentment or frustration to such individuals, for these notions presuppose that the person possesses a level of conceptual function which is absent or, at best, unavailable to observation. Even if we had ample material information, it is not clear that we could judge it according to the same standards by which we evaluate normal individuals. As John Robertson notes in his discussion of severely impaired newborns, it is inappropriate to judge these infants' lives according to a "standard based on healthy, ordinary development .... One who has never known the pleasure of mental operation, ambulation and social interaction surely does not suffer from their loss as much as one who has" ([37], p. 254; also, [31], pp. 13—14.)

These obstacles to describing and evaluating the quality of life of completely incompetent individuals account for much of the consternation experienced in any attempt to identify our moral responsibilities toward these individuals and subsequently to construct appropriate public policies. Since it is not possible to determine with confidence what these persons' interests are, the moral task is not the usual one of determining how best to serve their interests. Rather, it is the task of managing substantial uncertainty: how we ought to treat these people, given that we really do not know what benefits them or what harms them — or, in some cases, whether these concepts can even be predicated of them.

Thus far I have spoken only of those who have devastating impairments of competence. With individuals whose competence is more intact, it may be easier to construct plausible objective, if not subjective, quality of life judgments. Again, however, there are problems. A particular individual's conceptual and linguistic deficiencies might preclude understanding his subjective point of view and this, in turn, may mean that at least some objective standards of quality of life are not entirely applicable. A physical handicap which poses enormous frustration for a person of normal mentality, for example, may have no such effect on someone with very limited intelligence. To the extent that the usual objective evaluation of the handicap is based on its presumed adverse psychological effects, that evaluation may be of limited application in these special cases. In sum, the further an individual's autonomy is impaired, the less we are able to identify his subjective quality of life or to assess even his objective quality of life.

IV. THE RELATIONSHIP BETWEEN OBJECTIVE AND SUBJECTIVE QUALITY OF LIFE

It should be clear from the foregoing discussion that it is inappropriate to speak of the quality of life simpliciter. The concept is ambiguous and requires division at least into objective versus subjective standards. Our shared community standards reflect basic societal values. They identify the things for which, collectively, we have our greatest fears, aspirations, and sympathies. These must not be equated with the fears and hopes of the particular individuals who personally face the situations which the community at large usually confronts only *ex hypothesi.*

Consequently, it is entirely possible in any given instance that a person's objective quality of life may conflict with his subjective quality of life — and yet that each evaluation be equally correct in its own right. Objectively, it is a bad thing to have a painful, fatal illness. Yet in a given instance, the person afflicted with such a disease might find in his affliction an opportunity to reconsider the meaning of his life and to make his remaining days a richer experience than he might ever have had otherwise. Individuals can manage to adapt successfully even to the most clearly undesirable circumstances ([15], p. 557).

At the same time, the two senses of quality of life are not utterly disparate. They can, and sometimes should, influence one another. Our collective judgments regarding particular states of affairs — for example, the quality of life we associate with a handicap such as paraplegia — are in part based on beliefs about what it would be like, subjectively, to live with such a handicap. To a certain extent, these beliefs are the reactions of healthy people as they imagine themselves to be in such a situation. And partly they are the product of afflicted persons' reports of their own subjective quality of life. If research were to show that the community's beliefs about the psychological impact of given predicament seriously diverged from the perceptions of persons actually in that predicament, then it would be reasonable to suggest amending the relevant objective quality of life judgment accordingly.

Reciprocally, subjective judgments about personal quality of life may be influenced significantly by the wider, community view of an individual's quality of life. A person who is elderly and poor, for example, may come to believe a societal judgment that he is unproductive, a drain on community resources, and thereby adopt a low appraisal of his own life ([5], p. 249). While such an influence sometimes may be inappropriate, there are times when it is right to expect a strong correlation between objective and subjective judgments. Sometimes the bare fact that a subjective judgment clashes with our objective quality of life standards can prompt us to question the validity of that subjective judgment or to examine the competence of the person who made it. If, for example, a person found an objectively minor problem such as a small facial scar to be a personally devastating affliction, it is reasonable to look for an acceptable rational explanation for this assessment. If the scar meant the end of a lucrative modeling career, then we can understand the glum judgment; if there were no such rational explanation, we might begin to question that person's emotional stability.

In sum, objective and subjective quality of life are distinct, albeit related, notions. As I will now suggest, each should have a role both in the formation of public policy and in the economic analyses whereby the achievements and efficiency of such policy are assessed.

V. PUBLIC POLICY AND QUALITY OF LIFE

This objective/subjective quality of life distinction has important implications for public policy and for the economic analyses with which such policies are evaluated. These implications can be best sorted out if a distinction is made between those public programs which are intended primarily to promote the welfare of society as a whole and those which are designed mainly to benefit citizens as individuals. Many social programs espouse both aims, of course. Nevertheless, the distinction will be useful in understanding the respective roles of objective and subjective quality of life.

Programs which promote the welfare of society as a whole would include, for example, national defense, environmental pollution standards, and any other program whose benefits are society-wide in scope and which generally are not achieved by distributing goods to individuals. Arguably, some public health programs can fit under this heading. These include sanitation, health information campaigns, or immunization programs which, although admittedly distributed to and of benefit to individuals, are intended for the benefit of the whole (and may be achieved at some cost to individuals, for example, in pharmacologic side-effects).

In these cases concern lies almost exclusively with the objective quality of life. Policy makers are engaged in determining basic societal directions — what sort of community will be promoted and how its future directions will be shaped. These directions involve the very value judgments which appear as the normative component of objective quality of life ascriptions. Such values identify the community's strongest and most agreed upon views of what counts as a good life, what is to be avoided, and what is hoped for. Arguably, policy makers have a moral obligation to appeal to such community standards as they design public policy. This is not to say that these standards are the only relevant consideration or that all policy should be a product of popularity or vote. The point here is simply that because the benefits (and costs) are community-wide, it is the standards of the community, and not those of particular individuals, which should carry special weight. Insofar as quality of life considerations are invoked, the objective quality of life is the relevant concept.

Matters are more complicated when considering those programs designed to serve individuals as well as the wider society, or those designed exclusively to serve individuals, that is, to help individuals for their own sakes regardless of whether wider benefits accrue to society. Programs to improve prenatal care and infant nutrition might exemplify the former situation, while institutions which care for people with severe retardation or advanced dementia exemplify the latter.

Some observers, including T. M. Scanlon, argue that here too we should appeal exclusively to objective standards of quality of life in formulating public policy [40]. If we did base our public assistance programs (or for that matter, our broader moral duties of mutual assistance) upon individuals' own tastes and interests — upon their subjective evaluations of their circumstances — we could find ourselves under obligation to support extravagant desires, inefficient uses of resources, or eccentric tastes simply on the basis of the subjective intensity of the individuals' desires. This, Scanlon suggests, is untenable. We do not bear obligations to support such preferences. Rather, "the criteria of well-being which we actually employ in making moral judgments" and in criticizing and justifying moral and political institutions "are objective", that is, independent of individuals' particular tastes and interests ([40], p. 658). It is at this point that Scanlon argues that only our most urgent, central interests are appropriate objects of public policy.

There is much to be said for Scanlon's position. Even in democratic societies in which citizens influence public policy and the level and uses of taxes, the fact remains that citizens' economic contributions to the nation and its policies are not altogether voluntary. Typically a citizen has no direct legislative vote and, except for impeachment or other extreme measures, his representative can only be replaced at designated intervals. To the extent that tax contributions are therefore conscripted, policy makers should ensure that only appropriate amounts are taken, for appropriate uses. Arguably, "appropriateness" here is to be defined at least partly according to the community's shared values concerning what constitutes a good society and a good life — that is, according to objective standards of quality of life. In general, people ought not to be forced to hand over their legitimately earned income to causes whose merits can be defended only by appeal to idiosyncratic preferences of individuals [14].

The matter is not settled, however, for there are powerful arguments on behalf of permitting subjective standards of quality of life to enter into the formulation of public policies designed to help citizens as individuals. In the first place, contra Scanlon, subjective quality of life should not be equated with mere strength of subjective preference. As was shown in Section III, subjective quality of life reflects the (competent) person's basic, and morally important, capacity for autonomously judging his circumstances according to his own chosen beliefs and values. To the extent that a public program aspires to help individuals for their own sakes rather than solely as a means to advance the commonwealth, those individuals' subjective perspectives cannot be ignored. It is those standards which largely determine, in a given case, what is actually to count as help.

Consider, for instance, the government's expenditure of nearly two billion dollars per annum for the care of patients with end-stage renal disease (ESRD) ([15], p. 553). If we measure such individuals' quality of life strictly according to objective standards such as the patient's level of function or his ability to work at a job for pay, then these people will be judged to have a poor quality of life ([48], p. 7). If, however, we solicit these people's subjective judgments of quality of life, we may be surprised. Despite their adverse circumstances, many of these people are able to enjoy fully or nearly the same level of satisfaction in their lives as the population at large [15]. Similarly, the bare fact that someone is elderly or that he has chronic ailments may render his objective quality of life poor, yet this hardly constitutes sufficient reason to conclude that his subjective quality of life is also poor, or that he cannot find satisfaction in his life [3]. Public policies that would deny or substantially abridge these people's access to health care solely by appeal to their objective quality of life would, in effect, deny the fundamental fact that human health and health care are highly individual and deeply personal.

To the extent that a policy aims to benefit individuals for their own sake, then, it is not possible to determine what will constitute benefit without consulting the standards of the designated beneficiaries. This is not to say, of course, that individual tastes alone should dictate public policy. Rather, public policies which aim at benefiting individuals as such cannot hope to benefit individuals maximally without incorporating those individuals' own views.

Thus far, I have made two basic points in this section: first, that at some point, policy makers must appeal to objective standards of quality of life if they are to justify extracting money from citizens to pay for public health care; and second, that health care programs designed to help individuals must be permitted ample latitude for subjective standards of quality of life. While these two points emphasize markedly different moral and political values, they can be reconciled.

At a first level, it must be acknowledged that it is necessary to appeal to objective standards of quality of life in order to identify potential programs which are worthy of public funding. Objective quality of life, thus, offers a threshold concept — an eligibility criterion — identifying the needs and interests which merit public support. Without satisfying this test, it becomes difficult to justify the imposition of the necessary tax levies with their concomitant constraints on citizens' freedom and resources.

At a second level, however, it must also be acknowledged that policy makers may distinguish among the eligible projects by introducing other values, most especially those pertaining to the impact which the (proposed) program has on individual lives — as judged at least partly according to those individuals' own perspectives. Objective standards still have a substantial role to play at this level, of course, particularly with programs designed simultaneously for public as well as private benefit or with programs for helping incompetent persons whose subjective quality of life cannot be ascertained. Nevertheless, for programs intended to assist competent people, it seems essential at some point to invoke the designated beneficiaries' views concerning their needs and preferences.

An important implication of the foregoing point is that when we wish to design our policies so that they will "do the most good" (a guiding value for those who urge the use of cost-benefit and cost-effectiveness analyses in creating public policy) or when we wish to evaluate existing program results, we must unpack "doing good" according to the particular quality of life standards which are most appropriate to the situation. Selecting the right standard(s) depends in large part on what we hope to achieve through the program.

VI. ECONOMIC ANALYSIS AND QUALITY OF LIFE

The guiding aim both of cost-benefit and of cost-effectiveness analysis, most broadly construed, is to help ascertain what we have achieved or expect to achieve with a particular program or policy and, thereby, to determine whether societal goals are being pursued efficiently. As pointed out in the introduction of this paper, those goals often are understood as attempts to improve or to maintain the quality of life of society and its citizens. It should be clear by now that quality of life cannot be regarded as a univocal concept. Rather, societal goals and achievements can be viewed either from the wider perspective of the community as a whole or from the narrower perspective of the individual citizen. There are particular circumstances in which each perspective assumes special significance. Therefore, economic analysis must somehow be able to incorporate both ways of measuring quality of life.

Interestingly, this recognition seems to be absent from much cost-effectiveness and cost-benefit literature. I cannot hope to survey all the relevant literature here, and so will confine my discussion to the task of examining the notion of Quality Adjusted Life Years (QALY) as articulated by J. W. Vaupel [44], R. Zeckhauser and D. Shepherd [47], T. C. Schelling [41], and others. I intend only to show that, if such a concept as QALY is to be employed at all, it must be offered in a conceptually more complex form than currently available.

The QALY concept is introduced as a device for helping to identify and quantify benefits in evaluating a given health care program. It is not enough, Vaupel suggests, simply to count the lives saved by a particular program because, to be precise, lives are not really saved as such. They can only be prolonged for some particular length of time beyond the point at which they would otherwise have been lost. Neither, however, do we simply wish to count life-years saved, for life is more than bare vegetative functioning. It is one thing to salvage someone for another year of pain-wracked misery or coma and quite another to extend a year of happy, productive time. Therefore, Vaupel concludes, it makes sense to count benefits in terms of QALYs ([44], p. 75—76). Typically, one rates a year of full function as having a value of 1.00 and death as having a value of zero — although sometimes some living conditions are rated as worse than death [6]. Then, by various means one determines what fraction of that full value can be assigned to a year of life with a particular impairment. "Thus, according to this approach, a year of life on dialysis is worth only one third as much as a year with functioning kidneys; a year of life after mastectomy for breast cancer is worth only six months of 'normal' life" ([3], p. 1298).

Currently, the QALY concept is employed with varying levels of sophistication and thereby with varying conceptual problems. At the most rudimentary level, there sometimes is a complete failure to distinguish between objective and subjective standards of quality of life. In these cases, the standard invoked is closest to what has been called objective quality of life as the economist solicits life-quality ratings from the citizenry at large. D. L. Sacket and G. W. Torrance, for example, have attempted to show not only that it is possible to determine with reasonable accuracy the values of the general public regarding various health states, but that these public values can and should be taken into account by those who plan health programs [39]. Similarly, R. M. Kaplan and his colleagues have constructed an Index of Well-Being based on a two-year study of the views of randomly chosen individuals in the San Diego area [22]. They believe the Index to be a valid guide by which to quality-adjust life-expectancy. In neither case, however, is there any special attention devoted to the distinctive perspective held by the people who actually suffer a particular malady. Only the views of the general public are sought.

In practice we can see this objective quality of life approach at work, for example, in M. H. Boyle *et al.*'s economic evaluation of neonatal intensive care for very-low-birth-weight infants. In order to adjust life-years saved in terms of life-quality, Boyle and his associates interviewed a random sample of parents with schoolchildren living in the town whose neonatal services were being studied [6]. Participants were asked to rate a total of 960 different health states along four basic dimensions: physical function, role function, social and emotional function, and health problems. The resulting utility analysis was then used to adjust life-years of survival according to perceived quality – from 1.0 for a year of full health to −.39 for a year whose poor condition participants believed to be worse than death. While it would have been impossible to obtain the subjective views of the infants themselves, of course, it is noteworthy that the study failed to survey the views of parents of very-low-birth-weight infants. Other examples of such general public utility analyses are not difficult to find ([42], [48]).

There are two problems with this exclusive focus upon public, objective standards of quality of life. The first problem is that, often, the analyst does not even measure objective quality of life very well. In these cases there is commonly a failure to distinguish adequately between the two component dimensions of objective quality of life, namely the material facts of the life-condition(s) under study and the community's normative consensus concerning that life-condition. In such situations the analyst typically attempts to measure only the former element by counting the prevalence of the life-condition under study, and not to make the equally important measures of the direction and level of societal consensus. Thus, analysts such as Vaupel assume that lives below age sixty-five are of constant and roughly comparable quality and that, after sixty-five the quality of life diminishes to zero by age ninety ([44], pp. 76—77).[5] Vaupel has replaced a crucial item-for-study with a simplifying assumption which bypasses one of the most important questions to be investigated.

In other contexts, the cost-effectiveness analyst may recognize that the direction and intensity of community consensus is indeed an object of direct study, but may adopt poor techniques for assessing that consensus. For example, Vaupel measures the value to be placed on elderly life by soliciting Duke University undergraduates' opinions about the age to which they think they would like to live. While Vaupel concedes that the "sample is admittedly a very unrepresentative one", he nevertheless contends that the study does provide some evidence that elderly life is (objectively) of significantly poorer quality than younger life ([44], p. 76n). This position is difficult to defend. To suppose such a sample could constitute supportive evidence for conclusions concerning the views of society at large arguably strains any definition of adequacy in scientific methodology. Even Kaplan *et al.*'s more elaborate two-year study of a sample population in San Diego relies on the dubious assumption that these Californians' views are typical of the nation as a whole ([22], p. 485n).

The second and deeper problem with this approach, however, is that an exclusive focus on objective quality of life ignores entirely the crucial difference of perspective between those who suffer a particular malady and those who do not. I have discussed the significance of this point above. Although the objective quality of life of people with ESRD generally is thought to be poor ([48], p. 7n), Evans and others have found that these people "are able to adapt to very adverse circumstances, expressing satisfaction with their lives" ([15], p. 558).

Interestingly, some analysts who focus principally on objective quality of life do recognize that a failure to include subjective factors can be significant. In their efforts to include quality of life in a study of the cost-effectiveness of medical management of hypertension, for example, William Stason and Milton Weinstein note that the QALY factors to which medical researchers most commonly attend — morbid events and such medication side-effects as allergic reactions, gout, and anemia — may not capture some of the most serious detriments to everyday quality of life for patients. Weakness, fatigue, and depression are considerably more common pharmacologic side effects, and frequently are chronic ([42], p. 738). The authors therefore suggest that quality of life may need to be construed differently, and indirectly they hint at the difference between subjective and objective perspectives. The distinction is not made explicit in their discussion, however, and, in this study at least, they do not deal with the inadequacy of existing approaches for appraising the quality of life.

The important lesson from the foregoing is this. If a health program aims to help individuals, we cannot assess adequately its achievements — much less call the program a waste — without specifically seeking the views of those whom the program was designed to assist. Subjective quality of life must be a part of any adequate study.

Not all economic analyses ignore subjective quality of life, however. For example, in his discussion of the use of estrogen therapy for post-menopausal women, Weinstein incorporates such considerations. In his first-pass analysis he shows that estrogen therapy generally has little effect on life expectancy in either direction and that costs and savings tend to cancel each other ([45], p. 314). When quality of life ratings are subsequently introduced, however, the cost-benefit conclusions change markedly. Ultimately, he concludes, these quality factors — consisting principally in the subjective values assigned to symptomatic relief — will be decisive in determining the worth of estrogen therapy.

Of particular interest are Weinstein's comments concerning the ways in which one measures (subjective) quality of life and the approach which he implements for his own analysis. His quality of life adjustment factors are not the product of empirical research at all, but rather are values he selects "for illustrative purposes". Thus, he supposes that symptomatic relief will affect a life-year by a factor of .01; that hip fracture results in "disability equivalent to a loss of .05 quality-adjusted year for each remaining year of life", and so forth ([45], p. 311). In the case of this particular research project, Weinstein's substitution of assumption for empirical research is not really problematic for his conclusions. Since he has already established that the therapy does not affect length of life and is roughly cost-neutral, it matters little what particular values are entered into the quality of life component. Any notable improvement in quality of life is likely to render the therapy cost-effective. Nevertheless, it is not difficult to imagine other studies in which the particular values which one enters into the QALY position can make a significant difference in one's calculations. That is, there are plenty of instances in health care in which one must weigh extended survival against life-quality. For these cases, there is no substitute for empirical data.

The Weinstein estrogen study illustrates another problem. I pointed out earlier that it would be wrong to consider objective quality of life to the exclusion of subjective quality of life in assessing programs designed to help individuals. In the Weinstein study it is clear that, reciprocally, it can also be a mistake to focus exclusively on subjective quality of life when appraising the cost-effectiveness of public programs. Weinstein himself argues in his concluding discussion that, while the decision to use estrogen will undoubtedly remain an intensely personal one between patient and physician, nevertheless "in a world of limited resources, it is incumbent to elucidate the impact of such decisions at the societal level" ([45], p. 315). Accordingly, a central purpose of his and others' cost-effectiveness analyses is to ensure that society is expending its resources as it intends — whether that intent is to promote broad societal welfare or to help individuals to secure particular benefits. If such an analysis is to show that a particular program is cost-worthy for society, some reference to those societal purposes — to objective standards of quality of life — seems appropriate.

Weinstein has shown that estrogen therapy has little impact on length of life and that its use is essentially cost-neutral — neither positively worthwhile nor a serious waste — until quality of life factors are included. Without some inclusion of broader social values, his study can still be vulnerable to Scanlon's challenge that such therapy simply caters to idiosyncratic or eccentric preferences. Perhaps it is safe to assume that society would approve the particular quality of life benefits offered by estrogen therapy, and that Weinstein is safe to omit an examination of public opinion from his own study. Nevertheless, it is not difficult to recall health fads which neither lengthen nor shorten life, which are valuable in the eyes of their subscribers, and yet which could not qualify under community value standards as projects worthy of public support. Here, surely, one needs empirical evidence to ascertain whether a project is cost-effective from society's point of view, whether its benefits are of sufficient objective value to warrant public funding.

In the final analysis, subjective and objective standards of quality of life can both be of great significance for any economist who wishes to incorporate quality of life considerations into his analysis. These standards should be kept separate conceptually and not combined into some single assessment of "the" quality of life. Further, there is no substitute for well-conceived, carefully executed empirical research to determine individual and collective beliefs and values.

Of course, this is not to say that such empirical research is not itself beset by major conceptual and methodological problems.[6] Where economic research is markedly flawed by such problems, society must seriously inquire whether even to continue such research, and if so how to utilize its results. On the one hand is the view represented by Victor Fuchs, who argues that since there is no escape from making choices, neither do we "have an option between evaluating and not evaluating. The only option is whether to evaluate explicitly, systematically, and openly, as CBA/CEA forces us to do, or whether to evaluate implicitly, haphazardly, and secretly, as has been so often in the past" ([19], p. 936; also see [46], p. 717; and [2], p. 70). In stark contrast is the view of Jerry Avorn which warns that studies which are known to be seriously flawed — where, for example, they carry systematic biases against the elderly — should not be used. The

> ... alleged quantification is imprecise and so laden with value judgments that such numbers are useless .... Little comfort can be found in the standard argument that

decisions on resource allocation are made on the basis of 'gut feelings' all the time, and that these methods are an improvement over such imprecision. Approximate as gut feelings may be, they remain superior to delusional systems, no matter how detailed they are ([3], pp. 1299—1300).

Fortunately, this important debate need not be settled here. My aim is solely conceptual: to show that if economic analyses are undertaken, they should incorporate a more sophisticated notion of quality of life than is currently found.

Toward that end, one question remains. I have stated that cost-effectiveness analysis ought, at certain points, to appeal to subjective quality of life — to the views of the designated program beneficiaries who actually suffer the malady for which help is offered — if it is adequately to measure the effects of that program. And yet, one might argue, there are serious dangers in the attempt to quantify and aggregate personal judgments. If, for example, ESRD patients' views of the various therapies available for their disease — in-center hemodialysis, home hemodialysis, peritoneal dialysis, and transplant — were collected, the temptation might be to cite this aggregate preference alongside data about the therapies' respective costs to conclude that only the most popular, cost-efficient therapies ought to be offered. Thus, if peritoneal and home dialysis are better liked by patients than in-center dialysis, it might be concluded that we should try to place all patients on these "preferred" treatments. But surely, it could be argued, this would defeat the original purpose of soliciting these people's views *as individuals.* It makes little difference whether one's personal perspective is trampled by the views of the community at large or by the views of the smaller group of fellow sufferers. If the individual cannot tolerate or is unhappy with that most-efficient, collectively best-liked form of therapy, then the program will have failed him. When this point is added to the fact that, by definition, cost-effectiveness analysis must seek just such aggregate views, one might conclude on this view that subjective quality of life should not be included in cost-effectiveness analysis. While it might be a good idea in principle to look at individuals' views, this is not something that can be accomplished adequately when highly personal feelings must be merged into bottom line numbers.

My reply to this argument is that the appropriateness of aggregating subjective judgments of quality of life depends on the purpose for which the tally is to be used. We can distinguish three types of health policy which might aim, at least partly, at helping individuals for their own sake: (1) basic health care; (2) special assistance which goes beyond basic care; and (3) research.

With respect to basic health care, it would be inappropriate to use aggregate measures of quality of life to impose particular health interventions or therapies upon individual patients. To do so would violate the personal nature of health and health care and would intrude seriously upon the physician-patient relationship and upon patients' autonomy. It is possible, however, to use such tallies to help identify areas in which extra funding is needed, or occasionally even to support or rebut the idea that a particular form of health care is not basic or is a waste of resources. If research shows that a particular group of people whose objective quality of life is poor can nevertheless maintain a satisfactory subjective quality of life, then we have evidence that health care resources utilized by these people are not wasted.

At times and for a variety of reasons, a society may wish to devote special attention to a particular health problem. A disease may be especially feared, its sufferers may elicit special public sympathy, or its cure might have a particularly favorable impact on the nation's general well-being and productivity [1]. If we suppose that basic care by definition consists in those things which constitute the "decent minimum of health care" of which some philosophers have spoken (however the term decent minimum is to be explicated), then these "extras" represent society's opportunity to follow its own special interests. These programs are not designed to be imposed upon anyone, but simply offered as an extra benefit. Here, aggregate measures of subjective quality of life might help to ascertain which health problems cause the greatest personal suffering and to identify those projects which could help the most individuals.

Similar uses for aggregate measures of subjective quality of life can be found to aid in selection of research projects. While society's choice of research may be morally constrained by other values such as justice considerations, surely it can be at least relevant to ascertain which health problems generally cause the greatest suffering to their victims.

VII. CONCLUSION

There are good reasons for a society to employ both prospective and retrospective empirical studies in setting public policy. Prospectively, it is important to have some realistic notion regarding what a particular program is likely to achieve and at what cost to other community aims, so that society can prudently pursue its chosen goals efficiently. Retrospectively, some sort of accounting is important not only to ascertain what has been accomplished in order to guide future planning, but to offer an account to citizens of what has been done with their resources and to ascertain whether or not the overall impact of social programs contributes to a fair and just society. It is essential, then, to be able to compare the benefits and sacrifices of different people, to have a way of "measuring the equality or inequality of shares" ([40], p. 655).

Such an accounting does not always need to be precise. Sometimes society needs only to know that the money has been used for its intended purpose — for example, society may need only to know that the food intended for famine victims has reached them, but not to tally how many life-years were saved. At other times, a more precise accounting is needed; it is for these occasions that the tools of cost-benefit and cost-effectiveness analysis are appropriate.

In this paper I have not attempted to determine which tools should be used under which circumstances. At most I have demonstrated a more limited point. Where an economic analysis purports to incorporate quality of life considerations, "quality of life" must not be understood simplistically. One must ascertain whether he wishes to study objective or subjective standards of quality of life or both; this, in turn, depends upon the purpose of the program — whether it aims principally to benefit society at large or to help individuals for their own sakes, or both. Only when they encompass this conceptual sophistication can economic analyses begin to be considered adequate in quality of life appraisals.

University of Tennessee,
College of Medicine,
Memphis, Tennessee

NOTES

* The author gratefully acknowledges the editors of this volume, George Agich and Charles Begley, for their help and patience through earlier versions of this paper.

[1] While it can make sense to speak of the "quality of life" of a community rather than of an individual, my discussion here is restricted mainly to quality of life judgments as applied to individuals.

[2] Note that I am using 'society' in a rather broad sense, somewhat akin to nation in its scope, in order to distinguish the society as a whole from its various subcultures, which are marked by values and lifestyles not shared by the wider society. Thus, when I refer to the shared normative judgments found in objective quality of life, I refer to the basic values which unite members in a single society rather than to those peculiar values which differentiate the society into subgroups.

[3] For further discussion of competence, see [34].

[4] Perhaps we might challenge the assumption that such brain functions are essential for subjective experience — on metaphysical or epistemic, if not on scientific grounds. If somehow we did have reason to doubt our assumptions about the lack of mental activity in coma, then our ethical principles regarding treatment of patients in persistent coma might require reexamination. See [11], p. 124.

Unless we have good reason to change our minds on this subject, however, it seems sensible to assume with Brody that there is little chance that the permanently comatose have any mental life ([7], p. 86).

[5] Although Vaupel invokes this assumption with the recognition that it is "rough and simple", it does feed his conclusion that it is better to avert the deaths of young people than to prolong lives of the elderly. A more sensitive awareness of the range of positive (and of negative) quality of life throughout life might have led him to favor improving the quality of older persons' lives as well as saving younger lives.

[6] For discussion of epistemic and methodological problems of cost-benefit and cost-effectiveness analysis, see [2], pp. 48, 62, and 67n; [3]; [4]; [8]; [16], pp. 273 and 276; [19]; [24]; [25], p. 1951; [26], pp. 587 and 589; [30]; [32], pp. 688—89; [35]; [38]; [43]; [46]; [48], pp. 18—33, 35, and 37.

BIBLIOGRAPHY

[1] Aaron, H. J. and Schwartz, W. B.: 1984, *The Painful Prescription*, The Brookings Institution, Washington, D.C.

[2] Acton, J. P.: 1976, 'Measuring the Monetary Value of Lifesaving Programs', *Law and Contemporary Problems 40*, 46—72.

[3] Avorn, J.: 1984, 'Benefit and Cost Analysis in Geriatric Care', *New England Journal of Medicine 310*, 1294—1301.

[4] Bayles, M. D.: 1978, 'The Price of Life', *Ethics 80*, 20—34.

[5] Bloom, G.: 1984, 'Some Thoughts on the Value of Saving Lives', *Theoretical Medicine 5*, 241—251.

[6] Boyle, M. H., Torrance, G. W., Sinclair, J. C., and Horwood, S. P.: 1983, 'Economic Evaluation of Neonatal Intensive Care of Very-Low-Birth-Weight Infants', *New England Journal of Medicine 308*, 1330—1337.

[7] Brody, H.: 1981, *Ethical Decisions in Medicine*, 2nd edition, Little, Brown & Company, Boston.

[8] Bryant, G. D., and Norman, G. R.: 1980, 'Expressions of Probability: Words and Numbers', *New England Journal of Medicine 302*, 411.

[9] Daniels, N.: 1985, 'Fair Equality of Opportunity and Decent Minimums: a Reply to Buchanan', *Philosophy and Public Affairs 14*, 106–110.

[10] Devlin, P.: 1971, 'Morals and the Criminal Law', in R. A. Wasserstrom (ed.), *Morality and Law*, Wadsworth Publishing Company, Belmont, California.

[11] Dresser, R. S., and Boisaubin, E. V.: 1985, 'Ethics, Law and Nutritional Support', *Archives of Internal Medicine 145*, 122–124.

[12] Dworkin, G.: 1976, 'Autonomy and Behavior Control', *Hastings Center Report 6*, 23–28.

[13] Dworkin, G.: 1982, 'Autonomy and Informed Consent', in President's Commission for the Study of Ethical Problems in Medicine and Biomedical and Behavioral Research, *Making Health Care Decisions, Volume Three: Appendices: Studies on the Foundations of Informed Consent*, U. S. Government Printing Office, Washington, D. C., pp. 63–81.

[14] Engelhardt, H. T.: 1984, 'Allocating Scarce Medical Resources and the Availability of Organ Transplantation', *New England Journal of Medicine 311*, 66–71.

[15] Evans, R. W., *et al.*: 1985, 'The Quality of Life of Patients with End-Stage Renal Disease', *New England Journal of Medicine 312*, 553–559.

[16] Fein, R.: 1976, 'On Measuring Economic Benefits of Health Programs', in R. M. Veatch and R. Branson (eds.), *Ethics and Health Policy*, Ballinger Publishing Co., Cambridge, Massachusetts, pp. 261–287.

[17] Feinberg, J.: 1984, *Harm to Others*, Oxford University Press, New York.

[18] Foot, P.: 1977, 'Euthanasia', *Philosophy and Public Affairs 6*, 85–112.

[19] Fuchs, V. R.: 1980, 'What is CBA/CEA, and Why Are They Doing This to Us? *New England Journal of Medicine 303*, 937–938.

[20] Griffin, J.: 1979, 'Is Unhappiness Morally More Important?', *The Philosophical Quarterly 29*, 47–55.

[21] Gutman, R. A., *et al.*: 1981, 'Physical Activity and Employment Status of Patients on Maintenance Dialysis', *New England Journal of Medicine 304*, 309–313.

[22] Kaplan, R. M., *et al.*: 1976, 'Health Status: Types of Validity and the Index of Well-Being', *Health Services Research 11*, 478–507.

[23] Kavka, G S.: 1982, 'The Paradox of Future Individuals', *Philosophy and Public Affairs 11*, 93–112.

[24] Kenney, R. M.: 1981, 'Between Never and Always', *New England Journal of Medicine 305*, 1097–1098.

[25] Klarman, H. E.: 1967, 'Present Status of Cost-Benefit Analysis in the Health Field', *American Journal of Public Health 57*, 1948–1953.

[26] Klarman, H. E.: 1982, 'The Road to Cost-Effectiveness Analysis', *Milbank Memorial Fund Quarterly 60*, 585–603.

[27] Kleinig, J.: 1978, 'Crime and the Concept of Harm', *American Philosophical Quarterly 15*, 27–36.

[28] McNeil, B. J., *et al.*: 1978, 'Fallacy of the Five-Year Survival in Lung Cancer', *New England Journal of Medicine 299*, 1397–1401.

[29] McNeil, B. J., *et al.*: 1981, 'Tradeoffs Between Quality and Quantity of Life in Laryngeal Cancer', *New England Journal of Medicine 305*, 982–987.

[30] Menzel, P. T.: 1983, *Medical Costs, Moral Choices*, Yale University Press, New Haven.

[31] Mill, J. S.: 1957, *Utilitarianism*, O. Priest (ed.), Bobbs-Merrill Educational Publishing, Indianapolis.

[32] Mishan, E. J.: 1971, 'Evaluation of Life and Limb: A Theoretical Approach', *Journal of Political Economy 79*, 687—705.

[33] Mishan, E. J.: 1975, *Cost-Benefit Analysis*, 2nd edition, George Allen and Unwin Ltd, London.

[34] Morreim, E. H.: 1983, 'Three Concepts of Patient Competence', *Theoretical Medicine 4*, 231—251.

[35] Moses, L. E.: 1985, 'Statistical Concepts Fundamental to Investigations', *New England Journal of Medicine 312*, 890—897.

[36] Rice, D. P., and Cooper, B. S.: 1967, 'The Economic Value of Human Life', *American Journal of Public Health 57*, 1954—1966.

[37] Robertson, J.: 1975, 'Involuntary Euthanasia of Defective Newborns', *Stanford Law Review 27*, 251—261.

[38] Robertson, W. O.: 1983, 'Quantifying the Meanings of Words', *Journal of the American Medical Association 294*, 2631—2632.

[39] Sackett, D. L., and Torrance, G. W.: 1978, 'The Utility of Different Health States as Perceived by the General Public', *Journal of Chronic Disease 31*, 697—704.

[40] Scanlon, T. M.: 1975, 'Preference and Urgency', *The Journal of Philosophy 72*, 655—669.

[41] Schelling, T. C.: 1975, 'Procedures for Valuing Lives', *Public Policy 23*, 419—464.

[42] Stason, W. B., and Weinstein, M. C.: 1977, 'Allocation of Resources to Manage Hypertension', *New England Journal of Medicine 296*, 732—739.

[43] Toogood, J. H.: 1980, 'What Do We Mean by 'Usually'?', *Lancet 1*, 1094.

[44] Vaupel, J. W.: 1976, 'Early Death: An American Tragedy', *Law and Contemporary Problems, 40*, 73—117.

[45] Weinstein, M. C.: 1980, 'Estrogen Use in Postmenopausal Women — Costs, Risks, and Benefits', *New England Journal of Medicine 303*, 308—316.

[46] Weinstein, M. C., and Stason, W. B.: 1977, 'Foundations of Cost-Effectiveness Analysis for Health and Medical Practices', *New England Journal of Medicine 296*, 716—721.

[47] Zeckhauser, R.: 1975, 'Procedures for Valuing Lives', *Public Policy 23*, 419—464.

[48] Zeckhauser, R., and Shepard, D.: 1976, 'Where Now for Saving Lives?', *Law and Contemporary Problems 40*, 5—45.

PETER S. WENZ

CBA, UTILITARIANISM, AND RELIANCE UPON INTUITIONS

I. INTRODUCTION

In the present essay, I explore the relationship between cost-benefit analysis (CBA) and utilitarianism. I find that CBA receives its greatest theoretical support from utilitarianism. In fact, it is nearly identical to one form of that ethical theory and shares with utilitarianism the ideal of replacing reliance on intuitive judgments with reliance on calculative rationality. At the same time, however, CBA suffers from some of the weaknesses of utilitarianism, which render it unreliable as an analytic tool. While modifications to CBA address some of these weaknesses, they virtually abandon the original ideal of replacing intuitions with calculations, because the modifications can be justified and employed only through the use of intuitions.

After relating CBA conceptually to utilitarianism, I discuss problems confronting the use of CBA and utilitarianism in health policy analysis and consider adaptations that have been made or proposed to deal with these problems. The problems concern distributive justice, allowance for time preference, price valuing of resources, and the reliability of individual preferences. I then examine some of the modifications developed to deal with these problems, such as giving extra weight to the preferences of certain people, sensitivity analysis, and replacement of CBA by cost-effectiveness analysis (CEA).

II. CBA AND UTILITARIANISM

Utilitarianism and CBA extol the ideal of replacing intuitive judgments with judgments based on calculations. Jeremy Bentham, the first great expositor of utilitarianism, believed that moral decisions could be justified through what he called his "hedonic calculus", in which the pleasures and pains associated with different courses of action are calculated and compared [2]. The right action is the one yielding maximum net pleasure to those affected by it (or minimum net pain, when no available action yields net pleasure). CBA is similarly

G. J. Agich and C. E. Begley (eds.), The Price of Health, 71–89.

committed to calculative rationality. It is "an analytical technique that compares the costs of a project or technological application to the resultant benefits, with both costs and benefits expressed by the same measure. This measure is nearly always monetary" ([11], p. 169). The goal is the identification, through mathematical calculation, of investment projects which yield the greatest net benefits to society as a whole. As economist Kenneth Boulding points out: "the fundamental principle that we should count all costs, whether easily countable or not, and evaluate all rewards, however hard they are to evaluate, is one which emerges squarely out of economics . . . ." [3]

This commitment to calculation is coupled with a disparaging attitude toward qualitative judgments based on intuition. Boulding maintains that without quantitative, financial information "all evaluation is random selection by wild hunches" [3]. Other proponents of CBA maintain similarly that one should quantify costs and benefits "rather than rest content with vague qualitative judgments or personal hunches" ([20], p. 730). On the utilitarian side, Bentham claimed that his hedonic calculus was the only alternative to reliance upon (possibly) irrational traditions and/or the (possibly) irrational predilections of a ruling group [2]. Contemporary utilitarians agree that calculations should replace intuitions as the foundation for moral judgment. J. J. C. Smart writes that the "common moral consciousness is in part made up of superstitious elements, of morally bad elements, and of logically confused elements" ([23], p. 173). Peter Singer maintains "that all the particular moral judgments that we intuitively make are likely to derive from discarded religious systems, from warped views of sex and bodily functions, or from customs necessary for the survival of the group in social and economic circumstances that now lie in the distant past" ([22], p. 516). R. M. Hare, too, dismisses our moral intuitions as prejudices [9].

One way of avoiding reliance on intuitions is to adopt the goal of maximizing efficiency by utilizing the following decision-making procedure: identify the available courses of action and pursue the one that provides the greatest net benefits over costs. In the health field, these would be alternative ways of enhancing people's health or caring for the sick. They may range from increased funding of cardiac care units, increased funding of health-related education for young people, and stricter controls on benzene exposure in the workplace, to stricter controls on sulphur emissions from coal-fired furnaces. From the CBA perspective, policies affecting the health of the public are considered irrespective of the academic disciplines or professional fields from which they arise. So, policies in the medical field (cardiac care units) are considered alongside policies in the fields of education (health-related education for young people), occupational safety (benzene exposure in the workplace), and environmental studies (sulphur emissions from coal-fired furnaces). From the perspective explored in this essay, these can all be called health policies, because they are all policies that affect the health of the public.

The cost-benefit analyst, then, identifies the costs and benefits associated with each alternative, and calculates each program's net benefits (benefits minus costs). Employing the principle of economic efficiency ([1], pp. 15—17) requires that the policy or program chosen maximizes net benefit. Economic efficiency is achieved when goods and services of greatest net value are produced from a given supply of materials and labor. So, cost-benefit analysis considers the total net value of all goods and evils resulting from an action, regardless of how the goods and evils are distributed. As E. J. Mishan explains: "Instead of asking whether the owners of an enterprise will become better off by the firm's engaging in one activity rather than another, the economist [using CBA] asks whether society as a whole will become better off by undertaking it instead of a number of alternative projects" ([16], p. 13).

Is economic efficiency a goal worth pursuing? The answer to this question relates CBA to the ethical theory of uilitarianism. Utilitarianism is a general theory concerning the evaluation of human conduct. Human actions are right, and justified from the moral point of view, if and only if they result in maximum good. According to Jeremy Bentham [2], good experiences (pleasure, happiness, etc.) are, ultimately, the only good things in the world. All other good things, like health and wealth, are good because they promote good experiences. Conversely, bad experiences (pain, unhappiness, etc.) are the only intrinsic evils. All other bad things, like diseases and physical handicaps, are bad because they foster bad experiences of one kind or another.

Having thus defined good and evil, Bentham's utilitarian theory consists basically of the injunction to act always so as to maximize good (good experiences) and minimize evil (bad experiences). Bentham maintained that all other moral injunctions rest upon and derive their legitimacy from this basic utilitarian maxim. For example, why is it morally required that people respect the confidentiality of the doctor-patient relationship? If the confidentiality of the relationship were not respected, many people would not confide in their doctors, would not receive appropriate medical treatment, and would therefore experience unnecessary pain and suffering. Thus, the intuition favoring respect for the confidentiality of the doctor-patient relationship is explained and justified through a calculation showing that such respect maximizes human well-being.

Contemporary utilitarians such as R. M. Hare [9] and John Harsanyi [10] have abandoned Bentham's hedonistic (pleasure-oriented) interpretation of good and evil in favor of what is called preference utilitarianism. According to preference utilitarianism, the good is identified with the satisfaction of people's preferences, and the bad with the frustration of their desires. The difference between hedonistic and preference utilitarianism, and a major reason for abandoning the former in favor of the latter, is illustrated by the following case ([9], pp. 143–144 and [19], pp. 42–45).

Imagine that advances in robotics enable a small percentage of the population to do all the work that needs to be done in society [24]. Suppose that by this time advances in pharmaceuticals make it possible for the vast majority of people to be drugged through the water supply. Suppose that the drug does not impair people's health, though it renders them incapable of any constructive, planned, or sustained activity. It puts them in a stupor, but makes them happier than they could be without the drug's influence.

A utilitarian whose goal is to maximize human happiness would have to advocate under these conditions a program of drugging people, because the program maximizes happiness. Some people may prefer lower levels of happiness in order to have the adventure and challenge of trying to live life according to their own plans. But a consistent utilitarian who is dedicated to choosing policies that result in maximum happiness will have none of this. He will advocate forcing the unwilling to go on the drug plan – perhaps even to the extent of supporting the drug's being surreptitiously introduced into people's water supply.

Many people are repelled by this implication of hedonistic utilitarianism, and therefore identify "good" with the satisfaction of people's preferences, and bad with the frustration of their desires. If people do not prefer to live their lives in a happy stupor on drugs, then a utilitarian consistently dedicated to maximizing preference-satisfaction, unlike a utilitarian dedicated to maximizing pleasure or happiness, does not have to advocate and implement a policy of drugging people against their will. The goal of preference-satisfaction brings the dictates of utilitarianism into better accord with our normal ideas of what is right and wrong.

According to many people, preference-satisfaction utilitarianism is the best possible guide for human conduct. Because it is goal directed, it features the use of calculative rationality. People can use calculations, rather than tradition or personal feelings, when choosing a policy, rule, or course of action. The goal, satisfying people's preferences, shows respect for the diverse values and aims of people, since each individual is considered the ultimate authority concerning his own preferences ([10], p. 55). The common goal of maximizing people's preferences therefore involves respect for individual autonomy.

CBA can be supported by showing that it is simply a form of preference-satisfaction utilitarianism. According to CBA, preference-satisfaction should be measured by people's willingness to pay for things. In many contexts, willingness to pay is a commonly accepted measure of preference-satisfaction. For example, we normally consider it reasonable to determine what kinds of consumer items people want by how much they are willing to pay for them. We assume that, given a choice among items of many different kinds, the expenditure of limited funds on some things rather than on others indicates a greater desire for those things than for the others. We assume that the purchase of these things satisfies more preferences. Bad things are those that people are willing to pay to avoid. People pay for cars because they need or enjoy the use of a car, and they pay for collision insurance on a new car in order to avoid an uncompensated loss in the event of an accident.

Willingness to pay also facilitates relating one person's preferences to the preferences of others. The same monetary units can be applied to everyone's willingness to pay in order to determine the different levels of satisfaction that a certain good produces in different people. One person is willing to pay $10,000 for a new car, whereas another is willing to pay only $5,000, and therefore settles for a late model used car. All other things being equal, we can assume that having a new car was more important to the first person than to the second.

When preference-satisfaction is identified with willingness to pay, CBA is equivalent to preference-satisfaction utilitarianism. Economic efficiency concerns the extent to which available resources are used to create things that people prefer. Efficiency is increased when goods and services of greater preferred value can be produced from a given resource base. Increases in efficiency, then, increase the supply of goods and services and thereby increase the ability of people to satisfy their preferences through voluntary purchases. When the satisfaction of such preferences is identified with the good, as it is by the version of preference-satisfaction utilitarianism under consideration, increases in economic efficiency are good, and CBA, which is aimed at maximizing economic efficiency, is identical to utilitarianism, which promotes the same goal. Thus, CBA gains justificatory support from its relationship to utilitarianism.

III. PROBLEMS IN APPLICATION OF CBA

Distributive Justice

Because CBA's goal is the maximization of net benefits *in society as a whole,* it, like utilitarianism, is liable to the objection that it endorses distributional arrangements that may be considered unjust. Consider a healthy woman with a rare blood type who wishes to play tennis professionally. Imagine that an unusually large percentage of people with this blood type are subject to a chronic illness which requires treatment that involves frequent blood transfusions. The problem is that because the individuals in ill-health are numerous, and their need for the blood of the tennis player (and of other healthy people with that blood type) is crucial, a utilitarian would have to endorse forcing the unwilling tennis player (and others) to give blood frequently. This is the only way to maximize total preference-satisfaction. The tennis player's preferences may not be satisfied. She may not like to give blood, and frequent blood donations may impair her tennis game, if not her health. But these frustrations for the tennis player are overridden by the satisfactions of those in ill-health. The critic of utilitarianism finds this implication to be troublesome, because the tennis player's rights are violated by these forced blood donations. She is not allowed to keep what is rightfully hers, her own blood. This is a form of stealing and, as such, involves distributional inequities. Since CBA is a form of preference-satisfaction utilitarianism, it is subject to the same objection.

One method of attempting to circumvent this problem involves appeal to the Pareto Criterion. According to that criterion, only those policies are acceptable which benefit at least one person and harm no one. Since no one is worse off as a result of the policy, no one has any grounds for complaint if the policy is adopted. Thus, when CBA is limited by the Pareto Criterion, it is immune to the objection that it supports unjust policies. In terms of my example, since the tennis player is harmed (at least professionally if not personally) by forced blood donations, the Pareto Criterion would rule out the policy of requiring such donations.

Unfortunately, the Pareto Criterion is too stringent to be practical. If CBA were limited by this principle, it would be of almost no use since

> there are very few policy proposals which do not impose some costs on some members of society. For example, a policy to curb pollution reduces the incomes and welfares of those who find it more profitable to pollute than to control their waste. The Pareto Criterion is not widely accepted by economists as a guide to policy. And it plays no role in what might be called "Mainstream" . . . economics ([7], pp. 8—9).

Instead of using the Pareto Criterion, the Kaldor-Hicks principle is applied ([12]; [14]). According to this principle, a policy that imposes costs on some people is acceptable so long as it is possible for those who gain from the policy to fully compensate those who lose. This is sometimes called the *net benefits approach,* because so long as a policy has net benefits greater than the status quo, it will (in principle) be possible for those who gain from the policy to fully compensate those who lose. It does not matter, according to the Kaldor-Hicks principle, whether or not the winners actually *do* compensate the losers. It is sufficient that they *could* do so. Thus, if the net sum of preference-satisfactions (as measured by people's willingness to pay for things) is increased by requiring frequent blood donations from all healthy individuals with a certain blood type, CBA argues for adoption of the policy requiring such blood donations. Since social wealth is thereby increased, the blood donors could, in principle, be fully compensated (monetarily) for their blood donations. According to the Kaldor-Hicks principle, the policy of requiring these donations is thus justified, even if no compensation is ever actually given to the blood donors. A similar justification could be given for a policy that allows people's vital organs to be used when they die. The policy need not include compensation for this use. An uncompensated kidney donation during one's lifetime could also be required. In short, when CBA is conjoined with the Kaldor-Hicks principle, it sanctions distributional inequities.

According to M. C. Weinstein and W. B. Stason ([26], p. 718), some CBA advocates believe that "over large numbers of programs and practices the inequities are likely to even themselves out and, with some exceptions, may reasonably be ignored". But, as Weinstein and Stason also point out, others take seriously the problem of inequities and recognize, for example, that if benefits are measured by people's willingness to pay for things, rich people's preferences will tend to have greater influence than the preferences of poor people in CBA. This result strikes many people as unjust, because CBA is supposed to give equal weight to the preferences of all individuals in society. Sensitive to this problem, R. Layard suggests that a dollar that a poor person is willing to pay be valued more highly than a dollar that a rich person is willing to pay [15]. This kind of modification might make the results of CBA more acceptable on distributional grounds, but how would it affect the relationship between CBA and utilitarianism? Would it disturb the support that CBA receives from utilitarianism?

Not at all, according to Hare, who writes:

> Almost always, if money or goods are taken away from someone who has a lot of them already, and given to someone who has little, total utility is increased, other things being equal . . . . Its ground is that the poor man will get more utility out of what he is given than the rich man from whom it is taken would have got . . . . Diminishing marginal utility is the firmest support for policies of progressive taxation of the rich and other egalitarian measures ([8], pp. 113—14).

Egalitarian measures must not be carried too far, however, lest societal wealth diminish as people lack the incentive of personal reward to motivate their productive efforts. Thus, according to Hare, utilitarianism endorses a more equal distribution of benefits than obtains in society at present, though some inequalities of benefit must be preserved to motivate the production of social wealth. The relationship between CBA and utilitarianism thus seems closer than ever.

A serious problem remains, however, because there is no formal, rational method for *calculating* the appropriate corrective to inequality. Exactly *how much* more should a poor person's dollar be valued than a rich persons? *How much* residual inequality of wealth is necessary in society to motivate people to be productive? There are no algorithms for addressing these questions, so recourse must be made to intuitive judgments. These are just the kinds of judgments that utilitarianism and CBA strive to make unnecessary. Thus, although CBA can be modified to incorporate distributional values, there is currently no formula for incorporating these values in mathematical calculations. As a consequence, equity considerations are often under-emphasized in practice.[1]

Discount Rates and Shadow Prices

In CBA future costs and benefits are discounted to account for differences in preferences caused by time. In general, the benefits (costs) of a project will be less (more) in the future because of a preference for current consumption over future consumption. One way to estimate the current monetary worth of a future benefit (or cost) is to calculate the amount that one would have to invest (or borrow) today at the discount rate in order to reap the benefit (or incur the cost) at the future time in question. The higher the discount rate and more distant the future time in question, the lower the current value of a future benefit (or cost).

The choice of discount rate often has a significant effect on the outcome of a cost-benefit analysis. Consider, for example, the different effects of using a 3 percent versus a 6 percent discount rate. Suppose the net benefits associated with a proposed new hospital structure are estimated to be $1 million in the forty-ninth year of the structure's use. The analyst using a 3 percent discount rate would discount three-fourths of this sum, and count the present value of that net benefit as only $250 thousand. At 6 percent, this same million dollar benefit has a present worth of only $62,500.

Projects whose benefits (or burdens) are long term will thus be evaluated differently, all other things being equal, when different discount rates are used. The 3 and 6 percent rates used in my example are within the range of rates which analysts often use.[2] An analyst who is positively disposed to a project can probably justify his preference through an analysis that uses a relatively low discount rate. An analyst who is hostile to the proposal can usually reach an opposite conclusion through an analysis that uses a much higher, though equally permissible, discount rate. Thus, the objective appearance of the analysis belies the fact that its conclusion may be decisively influenced by the policy preferences of the analyst or by the preferences of the analyst's employer. As with considerations of equity, the results of calculations may be influenced by intuitive judgments.

The same conclusion emerges from a consideration of the shadow prices used in CBA to value resources in monetary terms that represent people's willingness to pay. For resources traded in the marketplace, market prices are usually used. But many items are not available in the marketplace, and, therefore, have no market price. For example, public goods often have no market price, nor do very personal things, such as the affection of friends, pain and suffering, and life itself. Such things may be germane to a CBA, especially in the field of health. Levels of pain and suffering and the length of human life will often be affected by policies concerning public health. If CBA is used to arrive at and to defend such policies, monetary values must be assigned to these variables. Since there is no marketplace where these things are bought and sold, people's willingness to pay for them must be inferred indirectly from what they say or do. The process of assigning values in this way is called shadow pricing.

The uncertainty surrounding shadow prices is much greater than that surrounding discount rates, making the results of CBA even more liable to determination by the analyst's intuitions. Consider, for example, attempts to establish the value of a human life using the willingness to pay approach. Shadow prices have been established, for example, by considering the wage differentials actually required to induce acceptance of hazardous employment, people's willingness to pay for life insurance, and their responses to hypothetical questions about how much money they would demand in compensation for assuming certain levels of risk to their lives.

Each of these methods provides sharply different results. For example, when wage differentials actually required to induce people to accept hazardous employment were used to estimate the value of a human life, estimates varied from $136 thousand to $2.6 million ([6], pp. 187—188). The method of asking hypothetical questions is even worse in this regard with estimates ranging from $28 thousand to $5 million ([6], pp. 188—189). A more recent study of this kind by Andreas Muller and Thomas Reutzel has yielded results that are stranger still [17]. The implied value that respondents assigned to a statistical life (the life of no one in particular) was more than 100 times the value that they assigned to their *own* lives. The conclusion reached by Muller and Reutzel thus seems warranted:

. . . [A]lthough it is possible to develop willingness-to-pay questions which produce fairly reliable and apparently reasonable responses, such responses do not seem to be based on rational considerations consistent with the basic assumptions underlying cost benefit analysis . . . . [T]he willingness to pay responses seem to be more a reflection of guessing than of rational considerations ([17], p. 811).

Lacking a justified, objective measure of the monetary value of human life, analysts in the health field who employ CBA must simply *choose* a value for human life. The policies recommended by such analyses will often be affected more by this choice than by any other factor. The analysis will almost inevitably give a false appearance of objectivity and rationality to a conclusion which actually is based largely on intuitive and subjective grounds.

Sensitivity Analysis

One response to the difficulties noted above is to employ what economists call sensitivity analysis. Weinstein and Stason explain:

In this method, the most uncertain features and assumptions in the . . . calculation are varied one at a time over the range of possible values. If the conclusions do not change when a particular feature or assumption is varied, confidence in the conclusion is increased. If, instead, the basic conclusions are sensitive to variations in a particular feature or assumption, further research to learn more about that feature may be especially valuable ([26], p. 720).

The goal is to deal with subjective judgments in a way that avoids their dominating the cost-benefit calculation. But this is more than sensitivity analysis can realistically hope to achieve. One reason is the breadth of "the range of possible values" where equity considerations, discount rates, and shadow prices are concerned. Consider equity.

Some people think that wealthy individuals in the United States require further tax cuts to spur them to increased productivity. They advocate a flat tax. Others believe to the contrary that a more steeply progressive income tax is needed because the wealthiest ten percent of the population are already too wealthy compared to the poorest ten percent. The differences between these views are large enough to affect the results of some cost-benefit calculations that incorporate equity considerations. The equity debate is so persistent as to render unlikely the prospect of settling the issue through "further research to learn more about that feature".

The same points apply to controversies concerning discount rates and shadow prices. The "range of possible values" for discount rates is zero to ten percent ([26], p. 719). The range of values for a human life varies by more than two orders of magnitude. Variations of assumptions from one end of these broad ranges to the other are very likely to alter the results of one's calculations. "Further research" is unlikely to yield a timely solution to the problem, as debates on these matters seem far from conclusion. Worse yet, equity considerations, discount rates, and/or shadow prices are prominent in most CBAs, so problems concerning them cannot be dismissed as merely theoretical.

Another problem with sensitivity analysis is its prescription that "uncertain features and assumptions" be "varied one at a time". If the justification for sensitivity analysis is to deal with uncertainty by considering the effects on the cost-benefit calculation of the entire range of questionable assumptions, then assumptions must be varied *in combination* with one another. Varying them one at a time will not reveal the entire range of values that result from people making different assumptions about matters that are uncertain. The true range of possible differences of opinion is revealed only when the effect of each problematic assumption is multiplied by the effects of all the others.

In sum, when matters are genuinely uncertain, as they often are, CBA contains no device to free its conclusions from the undue influence of assumptions made and justified on the basis of intuition. The fact is, we do not have, nor is there the prospect of people soon obtaining, the kind of quantifiable data required to conduct cost-benefit analyses. For this reason the promise of CBA to free people from reliance upon intuitions is impossible to fulfill.

IV. COST-EFFECTIVENESS ANALYSIS

The conceptual and empirical difficulties discussed above have led some to adopt CEA, cost-effectiveness analysis, in place of CBA. CEA does not require that all values be expressed in monetary terms, so difficulties attending the assignment of monetary values (shadow prices) to non-market goods such as human lives are avoided completely. Instead, in the field of health care costs are measured in dollars, but benefits may be measured in terms of lives saved, life years, or quality-adjusted life years (QALYs). Again, Weinstein and Stason argue:

> The use of "quality-adjusted life years" has the advantage of incorporating changes in survival and morbidity in a single measure that reflects trade-offs between them. The ratio of costs to benefits, expressed as cost per year of life saved or cost per quality-adjusted year of life saved, becomes the cost-effectiveness measure ([26], p. 717).

With this evaluation criterion, the best policy is the one with the lowest cost per quality adjusted life year.

In theory, CEA is in a worse position than CBA concerning the replacement of intuitions by calculations. Within the health field all benefits may be translated into quality adjusted life years, whereas in education it may be level adjusted literacy years (the levels being the educational level at which a person is literate — ninth grade, eleventh grade, etc.). Quality adjusted life years, however, are incommensurable with level adjusted literacy years. There is no way to calculate that one quality adjusted life year is worth two or five or twenty level-adjusted literacy years. So intuitions or political processes must be used in place of calculations to compare the relative merits of programs for education versus those for public health, versus those for national defense, versus those for preserving the family farm. In theory, then, CEA diverges more than CBA from the ideal of using calculative rationality to indicate the most beneficial course of action.

In practice, however, CBA is not much better in this regard than CEA, due to divisions in administrative responsibilities in our society. CBAs are often commissioned by people whose responsibilities are confined, for example, to public health or, more likely, to some area within public health such as the control of infectious diseases. Programs within this area are compared with one another, but not with programs to improve the national parks, or to provide increased job opportunities in Appalachia. Ideally, these types of comparisons are what CBA is designed to make. But administrators who lack broad responsibility for the nation's welfare often limit CBA to compare only alternative courses of action within their area. So in practice, intuitions and political processes are used to establish priorities among competing areas of national concern. Calculations as such have relatively little influence.

V. IRRATIONAL DESIRES

Even when one can determine what people want, these wants are often a poor basis for public policy, because many preferences are artificial and/or irrational. To see the irrationality of measuring benefits and costs according to artificial desires, consider a case of post hypnotic suggestion. Imagine that a woman under hypnosis is told that after awaking from her trance she should untie and then re-tie her left shoe laces whenever the doorbell rings. If the suggestion works, she will behave as instructed. If you ask her while she is in the process of carrying out these instructions why she is untying and then re-tying her left shoe laces, the woman will respond with perfect sincerity that she feels like it or that the shoe laces felt too tight or too loose. It genuinely will be the case that she prefers to perform this behavior when the doorbell rings.

Since the behavior is preferred only because it was suggested to this woman under hypnosis, we do not consider it to represent her true self. If we can see that the behavior does not function to promote her interests or well-being, we are inclined to think that she would be better off without the influence of the suggestion. We are disinclined to aid her immediate preference-satisfaction by facilitating her fulfillment of the desire to untie and re-tie her shoe laces whenever the doorbell rings. We may refuse the next time she asks us for money to buy yet another pair of shoe laces. The artificial and irrational nature of her desire prompts us to discount her preferences somewhat. In situations like this, such discounting seems the only rational procedure.

CBA requires, however, that all preferences expressed in the form of people's willingness to pay for things be treated equally. This can yield strange results in the health field, because many people are willing to pay for the impairment of their health. Consider the widespread practice of smoking cigarettes. People smoke cigarettes because they perfer to do so. Advertisements associate the practice with good looks, sex, and success. The same associations were formerly promoted more subtly in the movies, where heroes from Bogart to Belmondo used smoking cigarettes as part of their trademark. The association of the desire to smoke with this kind of suggestion is reinforced by the fact that smoking actually is less functional than untying and re-tying one's left shoe laces whenever the doorbell rings. It is more expensive and is harmful to people's health. The desire for cigarettes seems irrational and calls into question the propriety of giving full weight to people's market preferences in matters concerning health.

This phenomenon is not confined to our culture. The Nestlé Corporation, in an infamous case, marketed infant formula to people in South America who could not afford to buy enough formula to feed their children properly, and who lacked water of sufficient purity to mix with it ([5], [18]). Their advertising campaign deprecated breast feeding as old fashioned and barbaric. Most people buying the formula, and their infants, would have been much better off with breast feeding, so the preference for the formula seems irrational. Yet CBA would have to give as much weight to the preference among such people for infant formula as to any other preference backed by an equal willingness to pay. Under these conditions, it does not seem rational for those dedicated to promoting public health to use CBA.

Another consideration shows that advocates of CBA have good reasons of their own to discount people's current preferences. People's willingness to pay is affected by previous public policies no less than by advertising. Consider, for example, how much you would be willing to pay for a servant who was literally your slave. I suspect that many people in our society today find slavery distasteful, due in part to a one hundred-twenty-year ban on slavery in the United States. The policy decision to prohibit slavery affects people's perception of slavery and, with it, their willingness to pay for a slave. Most people in the United States today, I suspect, would not be willing to pay anything for a slave, because owning a slave would clash too severely with their ethical ideals.

It may be hard to imagine it, but three hundred years ago, most people felt differently. At that time, because many people of means were willing to buy slaves, a proposal to abolish slavery would never have survived a cost-benefit analysis. Slaves might have preferred the abolition of slavery, but they had no money, and therefore could have no willingness to pay. From the CBA perspective, the only relevant people were abolitionists, slave owners, and people who wanted to buy (and use) slaves. A policy of continued slavery could have been justified on CBA grounds if the owners and those wanting to buy (and use) slaves were willing to pay more for the perpetuation of slavery than abolitionists were willing to pay for its abolition.

There are two lessons to be drawn from this example. First, it illustrates what was pointed out earlier about CBA and distributive justice: CBA cannot be relied upon to yield acceptable policy decisions when rights and justice are at issue.

The second point relates directly to the influence of past policy decisions on people's current willingness to pay. Many public policies that have influenced our current willingness to pay for things, such as the decision to abolish slavery, would not have survived a cost-benefit analysis. From the CBA perspective, then, our currently preferred expenditures have been shaped by illegitimate, irrational, or unjustified policy decisions made in the past. The untoward influence of these policies makes our *current* willingness to pay a poor guide *from the CBA perspective*. Currently preferred expenditures have to be considered suspect then, just as most of us consider preferences induced by hypnosis to be suspect.

This fact creates a problem for CBA. There is no particular means of determining which of our current preferences are tainted by unwise policy decisions of the past, nor can degrees of influence be readily established. Since the willingness to pay for one thing is conditioned by the willingness to pay for other things, any alteration of preferences for one thing will have a widespread effect on the demand for other things. Consequently, distortion of people's preferences resulting from unjustified policy decisions of the past may have a rippling effect across many or most preferences. CBA is thus in a bind. It calls for reliance on people's willingness to pay. Yet, even from the perspective of CBA, it is unclear whether people's current preferences are worthy of respect.

The generalization of these problems to any form of preference utilitarianism is straightforward. Whether or not preferences are measured by people's willingness to pay for things, it makes little sense to accord equal weight to preferences of equal strength when one preference is reasonable and the other is artificial or irrational. Also, many public policies in the past were not inspired by the preference utilitarian ideal. Lacking a demonstration to the contrary, it is reasonable to suppose that many policies which were not inspired by that ideal did not conform to that ideal. If our current preferences have been shaped by past policies that were irrational from the utilitarian perspective, and if our preferences would have been different if utilitarian policies had been followed instead, then our current preferences are irrational, and so a poor guide to policy, from the *preference-utilitarian perspective*. Thus, like CBA, preference-utilitarianism calls for reliance upon people's preferences, and at the same time includes considerations that undermine the justification for such reliance.

VI. CONCLUSION

Proponents of utilitarianism and CBA suggest that reliance upon intuitions in decision making is irrational. I have maintained that utilitarianism and CBA rely heavily upon intuitions, and are to that extent as irrational as the decision procedures that utilitarians and CBA proponents disparage. In a society such as ours, however, where people put special trust in quantified processes and results, the intuitive character of CBA is almost certain to be overlooked and possibly misunderstood. Even when warning labels are dutifully attached, the conclusions of CBA are almost certain to command a respect which they probably do not deserve. CBA will be respected as somehow more objective than conclusions reached through approaches that are avowedly based on intuitions. Certain intuitions, those of the analyst performing the CBA, will thus carry more weight than others in policy decisions, not because all intuitions were critically examined and some were preferred over others, but because the analyst's intuitions were subtly incorporated in the "objective" research.

The most rational procedure in these circumstances is to examine all views with a proper appreciation for the nature of their evidential support. Because CBA has a tendency to obscure for many people the nature of much relevant evidence, it tends to diminish rather than increase the role of rationality in decision making. I suggest that CBA, therefore, tends to mislead even when it is used in combination with other methods.

There is, nevertheless, a way in which considerations of costs and benefits can legitimately be used in health planning. Consider the model of the Supreme Court of the United States, which makes decisions in situations where the goal of protecting personal privacy or religious freedom conflicts with the goal of protecting public health or public security, or where the goal of protecting the integrity of the nuclear family conflicts with the goal of protecting the health of minor children. There are no clear rules for prioritizing these competing goals, nor is there a method of reducing them to a common denominator and assigning numerical values to different degrees of their achievement. So the answers to many normative questions raised by Supreme Court cases cannot be obtained through mathematical calculation. However, the Court opinions in many of these cases are certainly rational in the sense of weighing competing values and exercising human judgment in light of constitutional principles. It is this kind of rationality that I believe should be used to decide among competing goals and programs when people's health is at issue. I see no reason why the cost of achieving certain goals, and the costs of alternate ways of achieving the same goal, could not be incorporated into the consideration of policies that concern public health through a similar type of reasoning.

Sangamon State University,
Springfield, Illinois

NOTES

[1] For present purposes, I can endorse the conclusion concerning equity considerations of a Hastings Center Report:

> The traditional approach to CBA excludes formal consideration of distributional effects such as equity and fairness. Since economists disagree about how to solve this problem, equity considerations are likely to continue to be underemphasized in practice ([11], p. 175).

[2] No particular discount rate has gained general acceptance in the economic community. As Weinstein and Stason explain: "Currently, economists espouse discount rates, after correcting for inflation, of as high as 10 percent (the rate used by the United States Office of Management and Budget, subject to much criticism) or as low as 0 (or negative) percent; most concensus lies between 4 and 6 percent" ([26], p. 719).

BIBLIOGRAPHY

[1] Bailey, M. J.: 1980, *Reducing Risks to Life,* American Enterprise Institute, Washington, D.C.

[2] Bentham, J.: 1948, *The Principles of Morals and Legislation,* Hafner, New York.

[3] Boulding, K. E.: 1969, 'Economics as a Moral Science', *American Economic Review 59,* 1—12.

[4] Boyle, A. R.: 1983, *Acid Rain,* Nick Lyons, New York.

[5] Elliot, M. W. and Beauchamp, T. L.: 1983, 'Marketing Infant Formula', in T. L. Beauchamp (ed.), *Case Studies in Business, Society and Ethics,* Prentice-Hall, Englewood Cliffs, New Jersey, pp. 221—233.

[6] Freeman, A. M. III.: 1979, *The Benefits of Environmental Improvement,* John Hopkins University Press, Baltimore, Maryland.

[7] Freeman, A. M. III.: 1982, 'The Ethical Basis of the Economic View of the Environment', unpublished manuscript prepared for the Sixth Morris Colloquium, Boulder, Colorado.

[8] Hare, R. M.: 1980, 'Justice and Equality', in J. Sterba (ed.), *Justice*, Wadsworth, Belmont, California, pp. 105—119.
[9] Hare, R. M.: 1981, *Moral Thinking: Its Levels, Method and Point*, Oxford University Press, Oxford.
[10] Harsanyi, J.: 1977, 'Morality and the Theory of Rational Behavior', *Social Research 44*, 623—656.
[11] The Hastings Center: 1980, 'Appendix D: Values, Ethics, and CBA in Health Care', in Office of Technology Assessment, U.S. Congress, *The Implications of Cost-Effectiveness Analysis of Medical Technology*, U.S. Government Printing Office, Washington, D.C., pp. 168—182.
[12] Hicks, J. R.: 1939, 'The Foundation of Welfare Economics', *Economic Journal 49*, 696—712.
[13] Institute of Medicine: 1981, *Cost of Environment-Related Health Effects: A Plan for Continuing Study*, National Academy Press, Washington, D.C.
[14] Kaldor, N.: 1939, 'Welfare Propositions of Economics and Interpersonal Comparisons of Utility', *Economic Journal 49*, 549—552.
[15] Layard, R.: 1972, *Cost-Benefit Analysis*, Penguin Books, Middlesex, England.
[16] Mishan, E. J.: 1971, *Elements of Cost-Benefit Analysis*, Allen and Unwin, London.
[17] Muller, A. and Reutzel, T.: 1984, 'Willingness to Pay for Reduction in Fatality Risk', *American Journal of Public Health 74*, 808—812.
[18] Molander, E. A.: 1984, 'Abbott Laboratories Puts Restraints on Marketing Infant Formula in the Third World', in T. Donaldson (ed.), *Case Studies in Business Ethics*, Prentice-Hall, Englewood Cliffs, New Jersey, pp. 189—211.
[19] Nozick, R.: 1974, *Anarchy, State and Utopia*, Basic Books, New York.
[20] Prest, A. R., and Turvey, R.: 1965, 'Cost-Benefit Analysis: A Survey', *Economic Journal 75*, 683.
[21] Sagoff, M.: 1984, 'Ethics and Economics in Environmental Law', in T. Regan (ed.), *Earthbound*, Random House, New York, pp. 147—178.
[22] Singer, P.: 1974, 'Sidgwick and Reflective Equilibrium', *The Monist 58*, 490—517.
[23] Smart, J. J. C.: 1967, 'Extreme and Restricted Utilitarianism', in P. Foot (ed.), *Theories of Ethics*, Oxford University Press, New York.
[24] Vonnegut, K., Jr.: 1952, *Player Piano*, Dell, New York.
[25] Wenz, P. S.: *Environmental Justice*, unpublished manuscript.
[26] Weinstein, M. C. and Stason, W. B.: 1977, 'Foundations of Cost-Effectiveness Analysis for Health and Medical Practices', *New England Journal of Medicine 296*, 716—21.

PAUL T. MENZEL

PRIOR CONSENT AND VALUING LIFE

I. INTRODUCTION

People value life many ways. "Absolutely": Do I want to live? Do I think I ought to live? Relatively, viz-a-viz others: What are the relative values of your life and mine? Of an infant's life, a child's, a young adult's, a forty-year-old's, an octagenarian's? Relatively, viz-a-viz other things *in* life: Should I risk my life to climb this mountain? Should I use these resources to increase my chances of staying alive, or should I use them for other things instead? Should they be used to save you? As compensation to others if I die: For how much should I insure my life for my beneficiaries? What damages should be paid to my family if you have wrongfully caused my death?

The last several of these questions concern monetary valuations of life; the first do not. In any case, monetary valuation of life is absolutely indispensable in modern life. Money is essentially a society's most universal medium of exchange, that which by definition can be used by people to trade goods and shift resources to meet other needs. In this sense, then, life naturally takes on a kind of monetary value in modern society; given the many means available for saving and protecting life, the question: "Which of these life-saving or life-protecting measures should be supported and to what extent?" frequently arises.

During the last decade, discussion of this question by economists has been dominated by one model: so-called "willingness to pay." In this instance, the monetary value of a person's life is viewed as a direct function of the willingness of that person — and perhaps others — to pay for reducing the risks of loss of life. Elsewhere I have argued that this model, with some sharp qualifications, is essentially correct ([14], pp. 24—71). First, willingness to pay to reduce *risk* to life ought to be read as indicating a valuation of *life,* not just of freedom from risk. Second, willingness to pay *money* indicates a real *value* of life, despite the fact that without life money has no value. The value of life that such willingness indicates, however, is only a *monetary* value. Third, there is nothing inconsistent in making different monetary valuations of the

G. J. Agich and C. E. Begley (eds.), The Price of Health, 91–111.

same life in different circumstances — in low risk as compared to high risk situations, for example. These circumstantial variations in the monetary value of life only indicate the perfectly acceptable fact that the value at issue is a subjective one. Fourth, the basic values of maximizing overall welfare and respecting individual autonomy create a strong *prima facie* case for use of something like the willingness to pay approach. And fifth, basic principles of equality tolerate its use. There should be no objection to the model which, when properly understood, allows greater monetary values of life for the rich than for the poor. The model says that only the *monetary* value — not the overall value — of poorer persons' lives is lower than the monetary value of wealthier persons' lives. Furthermore, the monetary value of poor people's lives is only lower *for the poor persons* than the monetary value of rich persons' lives is *for the rich;* it is not "lower, absolutely" (whatever that might mean), or "lower for society". Consequently, it would be a gross corruption of the model to aggregate together the varying monetary life values of rich and poor.

In this essay I attempt a defense of the willingness-to-pay model based on reading willingness to pay as *prior consent to risk.* I argue that measures of the value of life based in willingness to pay ought to be used in health care contexts.

In the course of my defense I seldom refer to this method of valuing lives by its usual label, willingness to pay, or by the term for the more general process of which it is usually thought to be a part, "cost-benefit analysis". I prefer to speak instead of *prior consent to risk* — risk, that is, of death or ill health. Such phrasing hopefully not only will prevent being distracted by some of the most objectionable but wholly unnecessary features of the model, but will allow a more succinct view of what is its essentially correct kernel, prior consent.

Throughout the paper a basic concern will be the background question of whether it morally corrupts our respect for life to regard prior consent to risk as indicating a value of life useful for societal and institutional decision making. I consider this to be the same question, in more religious terminology, as whether monetarizing the value of life in this way profanes the sacred.

II. AN HISTORICAL NOTE

I begin rather circuitously, with an historical comparison. Scepticism that monetarizing life corrupts respect for life or profanes the sacred is not new. It is instructive, I think, to look at the same sort of reservation in the relatively recent past concerning the institution of life insurance. Such a comparison should help in understanding some of the moral reasons for looking at prior consent for the value of life.

Viviana Zelizer has traced the historical development of life insurance:

> Particularly, although not exclusively, during the first half of the 19th century, life insurance was felt to be sacrilegious because its ultimate function was to compensate the loss of a father and a husband with a check to his widow and orphans. Critics objected that this turned man's sacred life into an 'article of merchandise' .... Life insurance benefits ... became 'dirty money'.... ([20], pp. 596—598).

Resistance to life insurance in the early 1800s reflected general condemnation of any materialistic assessment of death and particular apprehension about commercial pacts whose fulfilment depended on death.

To meet this criticism, the marketers of life insurance increasingly came to depend on what was a very effective strategy: to avoid any talk that smacked of insurance's "profitable investment" and to speak instead of its moral, altruistic value. Sensibly and without mere advertising hype, they noted that purchasing life insurance would be the making of a gift, the sacrifice of personal consumable income in order to care for dependents. A sermon in the 1880s would later reflect this idea:

> When men think of their death they are apt to think of it only in connection with their spiritual welfare .... [But] it is meanly selfish for you to be so absorbed in heaven ... that you forget what is to become of your wife and children after you are dead .... It is a mean thing for you to go up to heaven while they go into the poorhouse ([20], p. 603).

Life insurance "can alleviate the pangs of the bereaved, cheer the heart of the widow, and dry the orphan's tears", were the words of an earlier 1860 sermon. "It will shed the halo of glory around the memory of him who has been gathered to the bosom of his Father" ([20], p. 602). The U.S. Life Insurance Company's 1850 booklet described it as "the unseen hand(s) of ... [earthly] provident father[s] reaching forth from the grave and still nourishing ... [their] offspring ..." ([20], p. 604). By the late 1800s life insurance had taken hold, and it then became possible to market life insurance in more explicitly economic terms. Far from profaning life and death, it had taken on ritual and symbolic functions despite its potential abuses.

One can, of course, be cynical about the motivations of the marketers. Because of that cynicism any sacralization of life insurance can never be complete. Even if an individual is morally blameless in buying insurance, and regardless of the personal motivations of those who sell it, the life insurance business still institutionalizes self-interest. As Zelizer concludes, "particularly when it comes to death, to save and to heal is holier than to sell .... Marketing death is [still] ... 'dirty work'" ([20], p. 607).

The general parallels to all this in contemporary health care are striking. If society derives limits on what to spend to save lives from the previous consent given to letting risks stand because they were too expensive to remove, is life profaned in the process? To be sure, if others greedily encourage individuals to run those risks, they will be like life insurance beneficiaries who first encourage an individual to take out an insurance policy and then smirk with anticipation of fortune at the thought of that individual's death. So also, the sale of a *deviously* lean, cheap health care plan by profiteers who cut costs at every corner points to health care, health insurance, and human life itself having been profaned. Just as with life insurance, though, it does not take much to recognize the containment of health care costs on the basis of prior consent to risk as noble, moral behavior. Cost containment is simply not allowing individuals to self-indulgently utilize an insurance pool's resources merely because they are not now paying out-of-pocket for them. Society thus benefits in having more resources to use on diverse things — others' lives and health, or any dimension of the quality of life. This is true even if the motivation for individuals' willingness to take risks is purely selfish — for example, to have more money to pay bills, to get the car repaired, or to take a vacation — because that may represent their honest assessment of their present life, their financial resources, and their source of personal satisfaction.

Thus, far from cheapening respect for life, consent to risk and the model of willingness to pay can represent the most altruistic and realistic orientation. To see it that way, however, some marginal cultural shift may be required (as was in the previous moral acceptance of life insurance), but to say that such cannot or will not happen is very unimaginative indeed. Consent to risk and refusal to pay may indicate an emphasis on quality of life for self and others that is the highest-minded personal and moral idealism.

So far the argument is one of possibility: prior consent to risk can represent both an altruistic and intelligently realistic attitude toward death and life-saving health care. The argument, however, must go further: What reason is there to think that prior consent does (or will) represent these virtues enough such that health care policy should be developed in its conceptual terms?

III. WRONGFUL DEATH AWARDS

To see the distinctive appropriateness of the concept of prior consent for health care policy, it is instructive to look at the monetary valuation of life in a different context: wrongful death awards. In fact, if there were any one phenomenon in modern society which already constitutes an explicit monetary valuation of life, it would be compensation of survivors for a victim's wrongful death. In any case, the legal context becomes important for gaining a complete picture of how society ought to value lives.[1]

The logic of compensating survivors is full of ironies, and the stubborn conviction that a person's life is sacred and irreplaceable leads to a tremendous range of wrongful death awards. On the one hand, since the victim cannot be replaced and money cannot make up the real loss, why award anything beyond the actual financial loss sustained by the survivors? If additional money just cannot make the survivors "whole" (as the lawyers say), why continue to act under the seeming illusion that it will? Yet there is another argument with precisely the opposite conclusion: that if the total personal value of the victim to the survivors were priceless, then extremely large awards should be made, with no thought as to the limit of the awards allowed. After all, since the life that is now gone is priceless, it could never be said that any particular award, no matter how high, would make the plaintiffs more than whole.

In fact, the wide range of potential award can be explained by assuming, perhaps grudgingly, that life does have a kind of *surrogate* monetary worth for survivors. *Actual* financial losses caused by a death, of course, can be calculated with all the sophisticated wiles and exacting ways of contemporary economists — the victim's discounted future earnings, for example, minus personal consumption.[2] That figure will often be upwards of half a million dollars. If that sum were all that was allowed in the award, however, the monetary value of a life would be calculated only in a very narrow sense — the actual economic *costs* of the death. If, of course, there was concern that putting any further monetary value on life was misguided or morally corruptive, the legal system could simply not award survivors anything more. As a matter of fact, though, courts have been allowing — probably increasingly — recoveries for intangibles. The broadest of these is *consortium*: "society", guidance, companionship, felicity, and sexual relations. In turn, "society" denotes the broad range of benefits each family member would have received from the other's continued existence, including love, affection, care, comfort, and protection.[3] These are different from the direct pain and anxiety to survivors caused by the death and its manner. The latter pain and anxiety do not directly refer, as consortium does, to the earlier positive value of the person's life now lost to the survivors.[4] Total consortium awards up to $500,000 for all survivors are not unusual.

These losses, though intangible, are clearly real and so there is a very strong *prima facie* case for including them in compensation. The question, of course, is how one could possibly assign a monetary value to such intangibles. Conceptually, in compensating survivors, what is being done when such estimates are made? It can hardly be that the cost of purchasing *replaced* consortium is being estimated. Generally, that just cannot be bought, and if we are sometimes tempted to think that it can be, we have usually been deluded into mistaking something short of the real thing. Perhaps instead only the cost required to obtain for survivors *satisfactions equal in amount or degree* to the satisfactions they had in consortium are being estimated. While this suggestion may be on the right track, there are still imponderables: What satisfactions of one kind are equivalent in degree to other satisfactions of another kind? Jury members can try to make their own judgments on that point, but in the context of losing consortium with a *particular* loved one, it becomes a more complex matter. It is doubtful that people in a countroom setting after the damage is done can come to recognize what the real trade-offs are; or, even if they can recognize them, it is doubtful that they are psychologically capable of making any actual judgment of monetary equivalence.

Given this dilemma, what options are available to tort law? It could refer back to some prior risk decision, say, the amount survivors would have been willing to pay beforehand to reduce a given risk to their consortium with the victim. That is surely a choice that juries sometimes make. Such a choice is certainly helpful in answering the question about what amount of money would compensate survivors for the loss of a loved one.

For lack of any better way of determining how much to award survivors so as to make them "whole", it might be a good idea to consider what monetary values are implied by people's prior consent to risk. Yet this move must be described as a very last resort for tort law. It does not and should not come easily. Tort law aims to make either victims or their survivors "whole". *Compensation for actual damage is the aim.* Just because a survivor might have taken a certain risk of the victim's death to retain something else that does have determinate monetary value does not by itself constitute any argument that some dollar figure computed from that willingness is all that a survivor should now receive. Yet in the absence of any other method of arriving at a monetary figure, and given the insistence of our convictions of fairness that survivors must be compensated with *something* for clearly real though intangible losses of consortium, what else can intelligent judges and juries do? To be sure, if an award conceptually derived from prior consent is taken to constitute *literally* the *actual* loss, it will profane the personal value of the victim to survivors. The legal system will best avoid this consequence by not instructing juries in any formal method that relies on prior consent to risk, but instead allow them to agonize; such conflict should frankly tempt them either to the one exteme of no award for consortium at all in order to proclaim that the lost life is priceless or to the opposite extreme of a very high award for intangibles to represent the infinite value of the victim.

Awards for wrongful death, then, ought to include essentially three components: pecuniary losses (the directly economic costs of death); particular disruptions created by death that do not directly represent the positive value that the deceased person had for survivors (for example, immediate pain and suffering); and the loss of the real but non-pecuniary value that the victim had for the survivors. Thus, if directly economic losses associated with death are realistically (as they often are) half a million dollars, and if immediate pain and suffering damages in a particular case were minimal, then most of the remaining half of a not uncommon $1 million award could stand for the personal value of the victim to the survivors.

It is interesting to note that a figure in this approximate range fits remarkably well with the results of the basic willingness to pay model for valuing a person's life. If one's prior consent to risk in order to save money generates dollar figures that gravitate around $1 million,[5] and if *other* people's valuations of that person's life are 40 to 50 percent of that individual's personal valuation ([9], p. 142), then wrongful death awards in precisely the range of $500,000 for the likes of consortium can be expected.

It is also appropriate to question whether wrongful death awards ought to be confined strictly to a victim's value to *survivors*, excluding personal value to oneself. After all, the dead individual is not experiencing damage, or even if in some sense he could be, he is not the one to be compensated were the value of personal life to oneself included in the award. Juries understandably would have difficulty acting on the basis of this point: Why should the tortious defendant avoid paying for the biggest injury of all — the one to the victim? Even if the damage is to survivors, why should the defendant not have to pay for one thing that the survivors, too, care about so much: the value of the victim *to himself*.

Traditional tort law has several interesting — though not in the last analysis promising — ways of dealing with these points. One way is through punitive damages, although they have always been problematic as a component in awards: Why should any particular individuals, in this case the survivors, get what represents the society's moral condemnation of the defendant? In any case, "punitive" damages are somewhat off the mark in relation to the current concern, in that they are awarded as punishment of a wrong and so do not represent the value of the victim's life to himself. Another way that traditional tort law might represent the value of the victim to himself would be to view his life as "property" that can be passed on to inheritors like any of the victim's other property. That possibility, however, founders on the legally and morally problematic conception of one's life as property. In the common law generally not even a person's body, much less that person's life, is regarded as personal property,[6] so it would be legally inconsistent to award compensation to survivors for the value of the victim's life, as if the victim were passing his life on to his estate.

In summary, it is understandable that prior consent to risk is not used in tort law as the key to establishing the compensatory value of a person's life, but by the same token neither the approach of lost discounted future earnings approach nor the human capital approach is used. In effect, the courts do not rely on any one formal rational methodology for setting the amount of awards. Given the lack of any better way of arriving at a figure for monetary compensation for intangible personal values of life, the use of prior consent to monetarize the loss of intangibles is tolerated in tort law. At least it can be said that tort law harbors no arguments against utilizing measures of the value of life by reference to prior consent to risk made in other contexts where the focus is quite different from compensation. Interestingly, the range of actual death award figures meshes remarkably well with the consent to risk values that people place on their own lives.

IV. WHY PRIOR CONSENT FOR HEALTH CARE?

In contrast to the problems involved in setting a monetary value for compensating actual losses and making survivors legally "whole", setting monetary limits on the value of life in the context of life-saving health care can benefit more agreeably from prior consent to risk. Should $100,000 be spent to add another 50 percent chance of survival to an individual's low, existing odds of living another five years? Here the question is not one of compensation for actual losses at all. Nor has any wrong been done that can form the basis on which anything is owed to that individual, family members, or friends. Furthermore, in a private context, at least, the money for providing the treatment often comes from payments into an insurance pool. Thus the question is not at all the one asked in the context of tort law. The issue instead concerns the amount of money necessary for the insurance pool to meet all claims and remain solvent. Taking individuals seriously in this matter necessitates referral at some point to prior consent to risk to save money. If a person knowingly chooses a leaner plan beforehand in order to have more money for things other than health insurance, that person's consent to the subsequently somewhat greater financial and health risks has become an important part of the social process of funding medical care.

Realistically these risks fall into several categories at the point of insuring: not just explicit exclusions of a specific list of services, but subtle reductions in the general standard of care. Enrollment in a pre-paid plan implies acceptance of an institutional arrangement in which it is absolutely predictable that health care personnel no longer will have completely free rein to do everything possible for the improvement of the patient regardless of cost per statistical benefit. Enrollment in a plan with a reputation for being especially cost conscious implies acceptance of an even more clearly acknowledged, lower standard of care.

Now it is possible, of course, that even beforehand, people creating the pool will decide that they do not want their later care limited by what they now choose to pay into the pool. Instead of "pre-paid" basis for limiting what is later spent to save them, they could conceive the arrangement as a more open-ended one. Even at the time of enrollment they might want the plan to spend according to what they would later be willing to risk in crisis. The amount spent per life saved then would presumably be higher than it would be if spending were limited to what people were willing to risk beforehand.

Which of these conceptual models — pre-paid or open-ended — best represents what people are doing when they fund an insurance plan? Undoubtedly this is partly a function of how economically pressed the society sees itself to be. Particular cultural values also affect what people are doing when they insure. A vigorous commitment to prevention, for example, may not only minimize the later need for care, but may actually be associated with a naturally greater prior personal willingness to risk being left without maximal treatment should crisis later strike. Heavily influenced simply by the fact that they now see themselves actively doing something significant for their health, people may then be more likely to say, "Well, that'll just be the breaks". For taking active responsibility for their own health, they have already reaped a major benefit: the psychological room to accept with equanimity more risks and, of course, benefit from being relieved of the financial burdens of high health insurance premiums in the meantime.

These admittedly relative economic and cultural influences on the act of insuring should not be a distraction from the central, underlying point: regardless of where in this spectrum between pre-paid limits and open-ended spending particular people or particular cultures fall, the prior consent question is still peculiarly appropriate for health care. Even the insurance buyer who wants things to be left unrestricted at the point of later needs is making that judgment from a prior vantage point. In modern society that ought to be the point of responsibility for the broad outlines of cost containment policy. That it ought to be is derived, first, from the way most health care is financed — by insurance, not out of pocket — and second, from the increasing realization of how overwhelmingly influential is the prior acceptance of risk embodied in a choice of insurance arrangement on the delivery of later actual care. Prior subscription in a plan thus ought to be seen as getting closer and closer to constituting people's actual consent to the risks to which the plan exposes them.

There are, of course, some major problems with such a conclusion. People may not understand how influential on the delivery of later actual care their prior choice of insurance arrangement is, or they may have little idea of the risks or level of care and range of services to which they are subscribing. Furthermore, in the case of publicly funded plans, they do not directly make their own choice at the insuring level at all. Consideration of both of these problems requires addressing another question before any conceptual cue logically can be taken from prior consent: Is it *actual* consent that is morally required for restricting care by cost, or is it proper to *presume* consent which reasonably can be extrapolated from available data concerning what people are willing to risk to save money?

In the context of publicly funded medical care, at least, I would argue that society may — and should — be guided strictly by a reasonable estimate of what people would have accepted beforehand had they been given an informed and explicit choice. In a democracy, even though people have not actually consented to particular features of government, a significant consideration nonetheless is that they probably have not wanted to take up all the particular rules of government policy one by one in a process of explicit consent. Instead, they have consented to those decisions being made by legislative or administrative representatives. In turn, at least one normative theory of representation — in fact, the theory of representation to which they would most likely consent if they reflected about representation itself — is that representatives should attempt to make judgments about what their constituents *would* want if they had to consider the particular matters at issue explicitly. Surely it is a good bet to say that citizens want someone else to make those judgments in government policy making *rather* than having government immobilized for lack of actual specific consent. Specifically, do not patients want health care providers to make intelligent, good-faith presumptions about patients' prior consent to risk rather than throwing up their hands for lack of explicit direction, doing nothing to contain costs at the cost-quality margin?

Suppose someone objects: But that violates their autonomy — hypothetical consent won't do, real consent is required [15]. Then, the shoe can be put on the other foot. For others' decisions about personal care, why would an individual *not* want others to use what that individual *really would* consent to if there were a chance to benefit from letting the risks stand? The alternative — doing little or nothing to contain costs, or something not reasonably related to an individual's choices — forces us to recognize that others respect an individual's free choices more if they were to use prices set in terms of prior consent in designing care than if they did not. That is true even if the resultant policies and practices leave that individual fatally short of what may have been life-saving care.

Of course there is the natural fear that others might overestimate the risks which an individual would consent to let stand in order to save money. But even with this realization, would it be a good idea to ask them to dispense with price-of-life ceilings? Hardly. The appropriate response would be to focus as accurately as possible on the prices that an individual actually places on his life. If there is still suspicion that others would underestimate the dollar value an individual would place on his life, why shouldn't they merely have to adjust upward their impression of an individual's price of life when they use it to let an individual's risks stand? Working this out would still fit squarely with the view that it is permissible to use monetary values of life computed from individuals' willingness to take risks to save money.

In conclusion, the case for the conceptual use of prior consent to risk to limit health care costs is overwhelming. If actual consent were given in an act of insurance, the justification of cost containment easily emerges. Even if consent were not given, if people *would* have accepted risks because of the cost of reducing them, the risks should still be allowed to stand under the following conditions: (a) we did not deprive the individuals of actual consent opportunities; (b) they share in the benefits of not spending more; and (c) their presumed price of life has been adjusted to account for any underestimates.

V. WHAT SORT OF PRIOR CONSENT IS REQUIRED?

There is a further important question for both representation in general and health policy in particular: How rational must the consent be for use in institutional or public policy? Whether the consent be actual or presumed, does it have to be rational consent, or may it be irrational? Furthermore, what is irrational as distinct from either non-rational or rational consent? I am particularly interested in the moral limits on using presumed prior consent, but the question about rationality also concerns the moral acceptability of actual consent. In any case, it is vital to clarify these matters or willingness to pay will be morally problematic.

To answer these questions, three conditions of consent should be considered: first, the general *knowledge* of the facts, say, about health care and the health care economy; second, an accurate and vivid *imagination* of the events that one is risking by a given prior choice and its alternatives; and third, *attitudes toward risk* — so called risk aversion or attraction. The rational elements seem to be the first two — the picture of the most rational choice incorporates them in their most complete form. The third element is either a non-rational or an irrational element, and a good deal hangs on which it is determined to be.

In the third category is the interesting and often allegedly irrational combination of attitudes toward risk to which Kahneman and Tversky have called attention in their work ([10]; [11]). For example, they discuss Allais's paradox. A person prefers

(a) a 100 percent probability of gaining $3000
or (b) an 80 percent probability of gaining $4000,

yet at the same time prefers

(c) a 20 percent change of winning $4000
to (d) a 25 percent chance of winning $3000

The rational choice in terms of more arithmetical chances of dollar pay-off is (c) over (d) as the person chose, but also (b) over (a), as the person did not. It is obvious what is influencing the person's risk aversion in choosing (a) over (b): an extra margin of perceived value is added when some gain is thought to be certain. Because of that, in fact, one naturally begins to wonder whether it is really irrational at all to prefer (a) over (b).

In other examples, however, Kahneman and Tversky have probably put their finger on genuine irrationalities. For example, people object less to a loss of $1200 compared to $1100 than they do to a loss of $200 compared to $100. To them the latter $100 difference appears bigger than the former one, but in the reality of human welfare the principle of diminishing marginal utility tells us otherwise. Even more strikingly, people are usually not willing to save $4 on a $125 jacket by driving across town to buy it, but they are willing to drive just as far to save $4 on a $15 calculator ([10], [11], [12]). The juxtaposed choices in each of these last two pairs of cases seem fundamentally irrational. It is not just that people have those preferences which cannot be justified by their contribution to some other goal — if that were all, they would only be non-instrumental, intrinsic, and therefore non-rational preferences. In these examples, however, people are simply inconsistent, all things considered.

If people's consent is presumed, which of these three sorts of conditions and components should be included in a complete picture of the actual person's hypothetical consent? It is not an easy question. The context is one of restricted resources requiring allocation decisions. Most people have given consent to someone else making those decisions for them by presuming what their consent would be rather than tediously gathering their actual preferences on each particular matter. Do people want their representatives to try to read their own consent as: (1) consent with perfect knowledge; (2) consent with perfect imagination; and (3) consent with (a) no irrational or (b) no non-rational attitudes toward risk?

It seems that on the first score, people usually want to be presumed to have most, but not all relevant knowledge — most in order to gain the benefits of knowledge, but not all because not even knowledgeable people in the relevant fields can claim that distinction. On the second score, people usually want to be presumed to have nearly perfect imagination — it probably fits better with common attitudes towards being highly imaginative than perfectly knowledgeable. As for attitudes toward risk, I would argue that people would want attitudes carried into presumed consent, as long as those attitudes are not irrational but only non-rational.[7] The reason is essentially simple: we can hardly eliminate non-rational preferences without robbing ourselves of most if not all of our very nature as desiring beings.

What remains to be argued is that the difference in implied monetary values of life in health care between prior consent, relatively low risk situations as compared to later, high risk ones is only non-rational, not irrational. There is the argument that the lower life-value preferences expressed in low risk situations are inconsistent with the goal of maximizing expected utility, but that would lead to a conclusion of irrationality only by blatantly begging other questions: Should attitudes toward risk be taken as constituents of 'expected utility'? Even if they should not be so taken, should maximizing expected utility to the exclusion of respecting autonomy of consent always be our dominant goal?[8]

A specific, provisional conclusion can be drawn from this argument: the presumed prior consent to a more restricted set of health care services, to lower standards of care, and in general to a lower monetary valuation of life that is due to the lower risk vantage point from which one makes a health insurance decision is a morally legitimate and not an irrational foundation for containing the costs of health care. To this point can be added a more general conclusion: while prior consent to risk is not a general theory of the monetary valuation of life, it is the appropriate stance for almost all purposes of putting a price on life in health care.

There is, however, one important qualification. Given the argument for this latter conclusion, one — essentially only one — category of exceptions inherently requires that the use of prior consent in health care be limited: cases in which people have never had a chance to benefit from such prior consent to risk — where congenital illness, for example, has always held them in a position from which life-saving health care seemed to be the most urgent and clearly first priority. Even in these cases, however, it is not that prior consent is the wrong cue; rather, it is just that prior consent in these cases does not yield anything like the clearly finite and relatively modest monetary valuations of life that emerge from most people's prior consent. From whatever vantage point in their lives, it cannot be presumed that they would be nearly as willing to take the risk of being without some beneficial health care as we can correctly presume that other people would be willing to take.

VI. AGE AND DISEASE FREQUENCY

Both this qualification and the general appropriateness of prior consent in health care can be put to particularly productive use in analyzing how health care resources should be allocated among patients of different age groups and population disease frequencies.

Consider age. From the standpoint of immediate or current consent, it is inappropriate to argue for a policy of spending less to save a seventy-five-year-old for possibly five more years than to save a twenty-five-year-old for possibly fifty more years. Assuming that this example involves a relatively vigorous and high quality five years of life for the older person, he probably now desires not to die nearly as strongly as the younger patient does. That probability should incline society toward spending nearly as much to save the old as the young, once the kinds of severe detractions from quality of life that can lessen one's desire to live are discounted.

But that is not the end of the matter. From a *prior* perspective, people undoubtedly would consent to a policy much closer than this one to maximizing the number of *years* of life, not just lives, that can be saved with health care dollars. Looked at ahead of time, the way to maximize chances in life is to consent to be bound by a policy much closer to maximizing the years of life which can be saved. After all, if a twenty-five-year-old dies because a seventy-five-year-old was saved with fewer dollars, or because more than one elderly patient was saved with the same amount of resources, the younger person will have lost much more than if he had turned out to be the older patient who died because the resources were used on the younger one. To be sure, because one must envision one's desire to live as a strapping seventy-five-year-old as nearly equal to one's desire when twenty-five, it would be inappropriate to consent to a policy that simply tries to save the most years of life, even though such consent will undoubtedly run a long way in that direction.

One of the reasons such prior consent seems not to be an unfair vantage point from which to look at the entire matter of age allocation is that every old person was previously young. That fact in itself does not imply that when everyone was young they naturally would have wanted to modify a simply-save-lives policy more in the direction of the young. If some people knew that they were much more likely than most to need lifesaving care when they were old and less likely than most to need it when young, they might indeed side with a policy of saving lives, not years of life. Barring that exception, it is thus remarkably easy to explain both why saving older persons for fewer future years of life should take lower priority, and why that is not age discrimination in any pejorative sense.[9]

Although the case of rare disease is more complicated, the moral analysis involved is also a productive example of my general conclusion. There are many reasons that I may have for being willing to take a risk in the future when I am paying for health care. The likelihood of my benefitting from a certain category of care even if I end up in crisis may be small enough so that, when combined with the probability of my ending up in crisis at all, I become willing to take a risk. On the other hand, the care in question, if I ever needed it, could be something which would very likely save me, and save me from otherwise certain death. But the odds of ever finding myself in that particular crisis may be so small that I would still be willing to take the risk of not providing that lifesaving care. A real example would be $200,000 liver transplants for liver failure.

There still may be a haunting suspicion that in consent to take risk, this latter sort of case is different from the former. In principle it is not, unless the rare disease at issue were one which all along in *my* life left *me* with a much higher than the average chance of someday needing the care in question. Am I permitted to discount my willingness to pay for the rarer one in proportion to my lower mathematical probability of ever needing the care? Why not? In relation to the role that prior consent should have in setting insurance coverage, I can find no relevent difference between disease frequency and other probabilities. For disease frequency, as for the other factors, the crucial cases in which prior consent does *not* work are when the *actual* people who subsequently incur the risk would *themselves* never really have consented to that risk at the earlier time. If, of course, a rare disease is early onset or congenital, prior consent thus would not work. Then, indeed, fairness would require that their care not be excluded from an insurance program just because others would give our prior consent to getting along without it.[10]

VII. THE TREACHEROUS USE OF EMPIRICAL DATA

An important question about presuming prior consent to risk in the context of health care concerns the risks that people in fact are willing to take. What are acceptable empirical data from which a presumption could be made of what people would risk? It is at this level, not at the level of basic moral justification, that the model of willingness to pay and prior consent to risk runs into its most significant obstacles to actual use.

In making a presumption of this sort, it is important to determine whether to refer to actual behavior or to stated preferences (questionnaires). This is not simply a controversy among social scientists about which vehicle will provide the most accurate data. It is necessary initially to decide what must be measured prior to any decision as to what can most accurately measure it. Does actual behavior constitute the morally relevant kind of consent that should be measured, or is it reflective, stated consent? Stated consent could be criticized because people may not really know what they want unless actual life choices put them to a final test of sincerity. On the other hand, most behavioral studies can be criticized on the grounds that it is doubtful that behavior actually reflects persons' genuine and informed desire, rather than other influences, including ignorance and powerlessness. Cost-benefit analysis and the use of prior consent to risk in policy decision making must remain sensitive to these problems.

Consider labor market studies. Suppose a study of high risk steel construction workers reveals that their consent to a higher risk occupation for a certain gain in income generates $300,000 in monetary value for their own lives. It may be argued that this value is unusually low, since people who take such risky jobs probably are less risk averse than others. On the other hand, in what segment of the labor market can be found the more typical consent to risk? Some common occupations combine rather low wage levels with surprisingly high risks of death. For example, waiters and waitresses in establishments where liquor accounts for the majority of revenue run some of the highest occupational risks of death, and bartenders are not far behind. Yet they have relatively low wage levels. Miners, surprisingly, run somewhat lower risks than they do, and roofers take below average occupational risks. Both of these latter occupations, however, have high wage levels.[11]

If many high risk occupations are associated with low wages, it is difficult indeed to read the labor market for behaviorally revealed preferences. An overview of what workers actually say about the risks they take presents evidence of inadequate information and frustration at the paucity of other current job options.[12] Then, either there will have to be accurate measurements of factors such as restricted job mobility and ignorance of risks affecting what jobs people are willing to take, in which case the labor market valuations of life could be adjusted intelligently, or appeal to some other means than labor market data will have to be made such as questionnaires or explicit rather than implicit market trades for safety — for example, seat belt or fire detector purchases.

The judgment thus is far from the price people actually do put on their own lives. Philosophical controversies lie at the bottom of the very choice between behaviorally revealed and explicitly stated preference. Each of these alternatives face major difficulties in application. Before those difficulties are resolved, however, it is still appropriate to follow intuition and focus on some roughly intermediate figure in the middle of the wide range of values implied from available data – something, say, like $1 million per life. The alternative – doing little or nothing to contain health care costs, or something not reasonably related to personal consent – is more of an insult to personal autonomy. If some compromise estimate for use in presuming consent is accepted, people will see themselves increasingly as having given prior consent in the act of intelligently choosing health care plans.

Pacific Lutheran University
Tacoma, Washington

NOTES

[1] Wrongful death awards may represent more than compensation of survivors for their losses: certainly prevention and possibly retribution are also general aims of tort law. I generally confine my discussion here to compensation and consider wrongful death awards only in their compensating component. I would argue that in fact this is the primary (the only?) *life-valuation* component in damage awards.

[2] This figure includes, of course, the market value of work that the person would have done for which payment was not received but which now either has to be done by the survivors or purchased by them on the market (e.g., child care or house and yard work).

[3] See, for example, *Elliott v. Willis* 422 NE 2d (1982, Illinois Supreme Court) and *Sealand Servi Gaudet* 414 U.S. 573 (1974, U.S. Supreme Court). In general see [1], [16], [17].

[4] See Blodgett [3]. For a very interesting discussion of some of the most basic theoretical puzzles about how to set commpensatory wrongful death awards, see Friedman [7].

[5] Baily's [2] composite figure (1980) is $350,000, while Smith's and Viscusi's [18] figures (1983) are several million dollars. For a review of these calculations, see Landesfeld and Seskin [13].

[6] The essential problem is that to own our bodies as property would be to call them things, but we conceptually associate our bodies too intimately with ourselves – persons – to do that. And *we* are not things, or at least we do not think we should be treated as things. The law has taken this point to heart. In some fascinating respects, courts have subtly negotiated control over our own bodily materials without calling our bodies property. In one case, for example, an unseeing eyeball was temporarily removed for examination for cancer and mistakenly dropped down an open sink drain

in the pathologist's lab. The plaintiff won on the basis of his psychological shock upon hearing of the bizarre accident, not on grounds of loss of property. See *Mokry v. University of Texas Health Science Center at Dallas* 529 S.W. 2d 802 (1975). In general see Dickens ([6], pp. 148—150).

[7] Underneath the surface this is an involved and touchy distinction. In its simplest form it is reflected in the comparative fact that you will probably spend virtually all the money you could possibly get your hands on to avoid an extremely high probability of dying but will be willing to spend *less* than a *proportionate* smaller sum to avoid a much lower risk. In other words, you imply a higher dollar value on your life by your willingness to spend your last million on one more year of life in the face of otherwise certain death than you imply by your willingness to spend only $10 to eliminate a 1-in-10,000 risk of death. See Menzel ([14], pp. 24—55).

[8] What is often regarded as the standard version of maximizing expected utility neglects attitudes toward risk — expected utility is simply the arithmetical product of the utility of a given outcome at the time it will occur and the presently estimated probability of that outcome coming about. Many have argued, however, that in fact the incorporation of present attitudes toward risk, not just the mathematical probabilities, should be used to adjust the utility of the outcome. In fact, Weirich has argued that the incorporation of attitudes toward risk into the calculation of expected utility should be regarded as the traditional, standard version of maximizing expected utility [19]. What is at issue here is simply whether prior consent should be the controlling consideration. There are no magical powers in the notion of 'expected utility' itself that can settle the moral question of whether prior consent and attitudes toward risk should be accounted for in whatever refined notion of expected utility finally is used.

[9] For a broader consideration of competing moral considerations on this matter, see Menzel ([14], pp. 186—194). On the matter of prudential, self-interested consent to lower priority for saving older lives, see Daniels [5].

[10] For a lengthier discussion, see Menzel ([14], pp. 203—212).

[11] For many surprising comparisons and an interesting explanatory account of such variations, see Graham and Shakow [8].

[12] See the numerous recorded testimonies in Brown and Nelkin [4]. For a very useful and constructive discussion of the problem of inadequate information, see Viscusi ([18], especially pp. 59—75).

BIBLIOGRAPHY

[1] *American Jurisprudence,* 2nd Series: 1977, *Proof of Facts,* Lawyers Cooperative Publishing Co., Rochester.

[2] Bailey, M. J.: 1980, *Reducing Risks to Life: Measurement of the Benefits,* American Enterprise Institute, Washington, D.C.

[3] Blodgett, N.: 1985, 'Hedonic Damages', *American Bar Association Journal 71,* 25ff.

[4] Brown, M. S. and Nelkin, D.: 1984, *Workers at Risk: Voices from the Workplace,* University of Chicago Press, Chicago.

[5] Daniels, N: 1983, 'Am I My Parents' Keeper?', in President's Commission for the Study of Ethical Problems in Medicine and Biomedical and Behavioral

Research, *Securing Access to Health Care,* Vol. 2, U.S. Government Printing Office, Washington, D.C.
[6] Dickens, B. M.: 1977, 'The Control of Living Body Materials', *University of Toronto Law Journal 27,* 142—198.
[7] Friedman, D.: 1982, 'What is "Fair Compensation" for Death or Injury?', *International Review of Law and Economics 2,* 81—93.
[8] Graham, J. and Shakow, D.: 1981, 'Risk and Reward: Hazard Pay for Workers', *Environment 23,* 14—20, 44.
[9] Jones-Lee, M. W.: 1976, *The Value of Life: An Economic Analysis,* University of Chicago Press, Chicago.
[10] Kahneman, D. and Tversky, A.: 1981, 'The Framing of Decisions and the Psychology of Choice', *Science 211,* 453—458.
[11] Kahneman, D. and Tversky, A.: 1982, 'The Psychology of Preferences', *Scientific American 246,* 160—173.
[12] Kahneman, D., Slovic, P. and Tversky, A. (eds.): 1982, *Judgment Under Uncertainty: Heuristics and Biases,* Cambridge University Press, Cambridge.
[13] Landefeld, J. S. and Seskin, E. P.: 1982, 'The Economic Value of Life: Linking Theory to Practice', *American Journal of Public Health 72,* 555—565.
[14] Menzel, P. T.: 1983, *Medical Cost, Moral Choices: A Philosophy of Health Care Economics,* Yale University Press, New Haven.
[15] Sagoff, M.: 1981, 'On Markets for Risk', Center for Philosophy and Public Policy, University of Maryland (unpublished).
[16] Speiser, S. M.: 1970, *Recovery for Wrongful Death: Economic Handbook,* Lawyers' Cooperative Publishing Company, Rochester.
[17] Speiser, S. M.: 1975, *Recovery for Wrongful Death,* 2nd ed., Lawyers Cooperative Publishing Company, Rochester.
[18] Viscusi, K. W.: 1983, *Risk by Choice: Regulating Health and Safety in the Workplace,* Harvard University Press, Cambridge.
[19] Weirich, Paul. 'Expected Utility and Risk' (unpublished).
[20] Zelizer, V. A.: 1978, 'Human Values and the Market: The Case of Life Insurance and Death in 19th-Century America', *American Journal of Sociology 84,* 591—610.

ROBERT AUDI

COST-BENEFIT ANALYSIS, MONETARY VALUE, AND MEDICAL DECISION

Medical decisions commonly have economic as well as moral importance. This holds whether they are about an individual or represent determinations of social policy. The moral issues surrounding medical decisions are complex and controversial, and in the past two decades they have been studied at great length by philosophers and others. The economic ramifications of these issues are also problematic and contentious, though on one thing there is wide agreement and some alarm: medical costs have been substantially increasing. The economic aspects of medical decisions raise a number of questions, and the pressures of rising costs, among other factors, have brought some of these questions to great prominence in philosophical discussions. Some of this discussion explores the applications of cost-benefit analysis to medical decisions. That cost-benefit analysis might be applied here is not surprising, given its apparent value in economics. There are, however, other reasons for its appeal. One major factor is the support such analysis apparently receives from decision-theoretic models as strategies for problem solving and rational choice. Another factor is the possibility of using data gained by cost-benefit analysis to ascertain individuals' values and thereby reflect them in medical decisions about those individuals. With these factors in mind, I shall explore some possible applications of cost-benefit analysis in the decision-theoretic tradition. Next, I shall assess some recent proposals for its use. My concluding discussions will suggest some new directions for research.

I. COST-BENEFIT ANALYSIS AS AN APPLICATION OF DECISION THEORY

Decision theory is now a well-developed branch of inquiry, and in the space available I can work only with a simple model: the maximization of expected utility model of rational choice. For the most part, this model seems to be the base from which more sophisticated models have been developed, and in any event it should serve my purpose. The central idea is generally familiar, and the following is one useful

G. J. Agich and C. E. Begley (eds.), The Price of Health, 113–131.

formulation. Where a person, S, can assign (subjective) probabilities to the outcomes he believes to be possible for each alternative action he supposes he can take, a rational choice for him, among these alternatives, would be (selecting) an action that maximizes his expected utility, that is, one which has at least as high a score as any alternative among the actions in question. Scores are computed by multiplying, for every envisaged alternative, the subjective value, for S, of each of its (subjectively) possible outcomes by the subjective probabilities of those outcomes, and then adding these products for each alternative action. Valuational numbers may be arbitrarily assigned as long as they are used consistently and preserve S's utility rankings.

Applying this formulation of the maximization model to a medical context, suppose Jan has two options, surgery and non-intervention. She might take surgery to have a probability of .60 of curing her, an outcome she values at, say, 100, and a likelihood of .40 of causing her death, an outcome she values at −75. She might take non-intervention to have a probability of .50 of resulting in cure, one of .20 of causing death, and one of .30 of yielding long-term partial remission, which she values at 30. The expected utilities, then, are, for surgery, $(.60 \times 100) + (.40 \times -75)$, i.e., 30; and, for non-intervention, $(.50 \times 100) + (.20 \times -75) + (.30 \times 30)$, i.e., 44. For her, then, the rational choice is non-intervention.

Clearly, this model is applicable in the pure economic domain: one need only assign values wholly in terms of money in order to have a method of applying cost-benefit analysis by the same strategy. Thus, if one could justifiably express certain medically relevant preferences in terms of monetary values, then cost-benefit analysis might help in at least some significant medical decisions. Suppose, for the sake of argument, that Joe knows that Jan would be willing to sell a kidney for $100,000. Imagine that she was injured, knocked unconscious, and, in the circumstances, could have her kidney saved only at a cost to her of much more than that. Then, other things being equal, if Joe must decide without consulting her whether to do the procedure, and he is attempting to do what is rational from her own point of view, he ought not to save the kidney.

There are many difficulties with the decision-theoretic model just sketched. Some of them are both sufficiently deep to be unavoidable by merely refining the formulation of the model and quite relevant to economic applications of the model to medical cases. First, there are many instances in which one simply does not have, and, on available information, *cannot justifiably form*, the appropriate range of probability beliefs: one for each of the usually enormous number of subjectively possible relevant outcomes. Second, even if one has enough probability beliefs, the model does not provide for discounting those beliefs that are either *irrational* (for instance, formed on the basis of sheer wishful thinking or an obviously unreliable source) or, on the other hand, simply false. Third, the model is similarly permissive regarding irrational valuations, though admittedly there is controversy over whether *intrinsic* valuations can be irrational. Intrinsic valuations are, roughly, valuations of something non-instrumentally and for its own sake. Since these are the kinds of valuations crucial for the model, it is important to ascertain whether they can be irrational and, if so, how the model should take account of that.

A fourth difficulty arises *if* one views the model as providing necessary and sufficient conditions not just for future or hypothetical cases, that is, for what choice (or action) is or would be *rational for S*, but also to actual conduct, that is, for *S's rationally choosing* (or acting). These notions are quite different, and as I have argued elsewhere,[1] the latter is subject to a broadly causal requirement according to which an actual choice (or action) is rational in the light of a reason, such as a set of values and probability beliefs, only if it is appropriately *based* on that reason. For instance, purely on the basis of superstition, rather than for any good reason, one might decline to swim in the ocean. In that case, even if one *has* reasons that would make declining rational *for* one (such as an aversion, which one has temporarily forgotten, to salt water), one does not *rationally do* it. The reason actually motivating the action does not make it rational; the reason that could make it rational does not motivate it.

Discussion of these difficulties would require a great deal of space and is not necessary here.[2] To simplify matters, I shall focus mainly on cost-benefit analysis in cases where these problems do not arise. There surely are some such cases, for instance where S is well-informed and adequately rational; and it would be significant if, even for that perhaps small set of actions, moral decisions in medicine could be aided by cost-benefit analysis. I want to explore this possibility in relation to some prima facie promising medical uses of cost-benefit analysis.

II. MONETARY VALUATION OF LIFE

If placing a monetary valuation on life itself is sometimes justified, then so, it would seem, is placing a monetary valuation on organs and on improvements in, or damages to, a person's health. Thus, monetary valuation of life is a good focus for assessment of some representative issues confronting cost-benefit analysis in medical decision making. How might such valuation be approached? It will not do to ask people what price they would name. For one thing, they are not selling their lives and thus would be at best speculating. But surely people do buy – and forego – insurance against specific hazards. They also pay to render their lives safer. It seems, then, that in a way the people in question are buying a reduced probability of certain undesirable outcomes. Why not suppose that the disvalue for them of those outcomes against which they take precautions, expressed monetarily, is the amount of money they spend times the reciprocal of the probability by which they consider themselves to reduce those outcomes? Thus, if Jan pays $10 to eliminate a 1-in-10,000 risk of death, the disvalue of death for her is apparently −$100,000; and implicitly, the monetary valuation she places on her life seems to be $100,000.

This reasoning must be qualified. As Paul Menzel says in refining his defense of the sort of view just sketched concerning how to arrive at the monetary valuation of one's own life, "you will probably spend virtually all the money you could possibly get your hands on to avoid an extremely high probability of dying but will be willing to spend *less* than a *proportionately* smaller sum to avoid a much lower risk."[3] One explanation for this disparity is aversion to risk. For those averse to risk, an *additional* disvalue accrues to choices they feel are risky. If the disvalue is sufficient to account for the relevant disparity, then one may be warranted in regarding the relevant choices as conforming to the decision-theoretic view of rationality and conclude that some choices which do not meet its standards are not irrational, but only *non-rational* and as such are still capable of guiding certain decisions affecting the person in question.[4]

There is a major problem here. Given how much people's valuations of their own lives apparently vary, if those valuations are to be crucial in medical decisions, then medical decisions that take account of the valuations must be informed by a way of eliciting, interpreting, and, in some cases, comparing them. A major part of the problem is to determine what conditions are most favorable to constructing a *morally valid* reading of people's valuations of their lives, that is, a reading from which it is morally permissible to draw conclusions about how they ought to be treated, say in making a medical decision. There is no simple answer to this problem. Menzel's conclusion is noteworthy:

> The judgment thus is far from the price people actually do put on their own lives. . . . [It] is still appropriate for us to follow intuition and focus on some roughly intermediate figure in the middle of the wide range of values implied from available data — something say, like $1 million per life. The alternative — doing little or nothing to contain health care costs, or something not reasonably related to personal consent — is more of an insult to autonomy.[5]

As Menzel's conclusion suggests, the main use he has in mind for monetary valuations of life is in containing the costs of health care. But exactly how might one ascertain and legitimately use such valuations? This is a difficult issue, but it must be resolved *before* monetary valuations of life may be used in reducing the costs of health care.

III. VALUATION, TACIT CONSENT, AND RESPECTING AUTONOMY

Regardless of whether rational choice may be defined by one or another decision-theoretic model, it remains an open question whether choice that is rational on the maximization of expected utility model is a suitable *moral* basis for conclusions about just how to regard the life of the person in question, even in the *prima facie* distant context of social policy decisions that may ultimately affect that person. An instructive comparison here is *informed consent.*

Defining informed consent is no easy matter. For the sake of argument, I take it as it normally is understood in the literature, namely, as at least free consent based on all the available relevant information about the medical procedure in question in relation to the subject's own welfare. Even if consent is both informed and maximizes expected utility, it may not morally validate a given procedure. Suppose, for instance, that it is heavily influenced by very low self-esteem, by a reckless willingness to take risks, or by irrational though true beliefs about the relevant probabilities (by good luck or because of manipulation, a belief might be true, and thus express information, even if it is irrational). S's informed consent may then fail to warrant a medical procedure.

The point is not that the maximization model fails to account for what constitutes rational choice — though that may well be so. The point is rather that the conditions under which a decision may be used as a basis for morally important choices affecting a person may well require *more* than just rationality and may certainly require more than rationality conceived as maximization expected utility. For one thing, there are surely degrees of rationality. Moreover, even if there were not, the conditions under which a rational choice is made may affect its status in presumptively justifying risky procedures to which S consents.

It might be natural to object here that since S has tacitly consented to a valuation on his life or on subjecting him to a risk of losing it, one would show disrespect of his *autonomy* if one did not behave accordingly. This objection might be pressed even if the consent does not appear unsatisfactory from one of the points of view just referred to, such as the moral point of view, which appears to set a standard different from maximization expected utility. This objection raises the question of where violations of autonomy end and warranted paternalism begins. No general answer to that question can be given here; but it is surely doubtful that S's autonomy is violated by one's refusing to regard risky surgery on him as warranted by consent he gives on the basis of low self-esteem and irrationally formed probability beliefs.

The point here does not imply that just *any* kind or degree of irrationality affecting consent entitles one to refuse to take the consent to warrant performing the procedure in question. The point is that some kinds and degrees of irrationality do invalidate consent (even in certain cases where it does not involve false beliefs about the relevant outcomes). It should be stressed that in refusing to take the consent as warranting the relevant procedure, one is not forcing S to do or undergo some procedure, but rather simply declining to regard an action of his as justifying something — such as major surgery — that clearly stands in need of justification. This distinction is crucial for understanding autonomy; the threshold for violating it is surely far *lower* for coercive procedures than for declining to perform one or another service which one is under no independent obligation to perform, but would give were one convinced that S had appropriately consented to it.

Quite apart from the difficult question of the nature of what might be called *warranting consent* — consent appropriate for moral validation of medical procedures of the sort that morally require consent — it is not clear just what sort of act constitutes *giving* consent, nor how consent is *transmitted* from something explicitly consented to, to something that is, in one or another way, implied by that. Thus, it might seem that if S has paid only the $10 mentioned to avoid a risk of one in 10,000 of death, then the *tacit* consent implied by this payment warrants reducing his safety by a medical decision which adds just a chance of two in 10,000 of death, yet saves him $25. But as reflection on this example suggests, the notion of tacit consent is far from clear.

To see the point about consent, compare belief. One can believe that Sue is Joe's brightest student and also that she is his most disappointing student, yet — particularly if one learns these things at different times — not believe something they evidently imply, namely, that Joe's brightest student is his most disappointing one. As is not uncommon, one fails to follow out a consequence of what one believes. In reply, it might be said that this point is of limited significance because one is at least *committed* to accepting what is evidently implied by what one already believes. That commitment, however, is only *prima facie*. For instance, on realizing that p implies q, one may be put in mind of reasons one has to *deny* q, and then, far from embracing q because believing p "requires" that, give up believing that p. If Tom, who is generally honest and reliable, says to me that his wife has exaggerated the distance between their house and mine, then ordinarily I may justifiably believe him and thus may justifiably believe that the distance is less than the 600 miles she said it was. But if I later discover that she was correct, or later recall confirming her estimate on the map, then far from being committed to believing her mistaken because I believe Tom, I should now give up my initial belief adopted on his testimony.

Why should consent be any different from belief in these respects? Indeed, since the moral justification of important decisions depends on consent, and since S needs not only to understand what he consents to, but to *consider it vividly*, the standard for *using* consent as a basis of medical decision should only in very special circumstances be any kind of tacit consent. There is an exception where the relevant consequence is either uncontroversially innocuous or so obvious that the original consent can be expected to extend to it. This exception normally holds for routine features S has already experienced in a kind of treatment, such as its requiring one's having one's temperature taken. But even here, it makes a great deal of difference whether what is regarded as tacitly consented to is something from which the patient can withdraw on finding it objectionable. That provision does not apply in the case just considered and in many others I shall discuss.

A related difficulty concerns *what is valued*, monetarily or otherwise. Valuing, even monetary valuing in the sense applicable to life, is not like *buying*. It does not have a physical object, or even an abstract but exchangeable object like a legal right. Valuing is a propositional attitude, and its object is something that is in some sense *envisaged* by a person, presumably a state of affairs conceived in a postive way. If the object is a state of affairs or anything else with the same degree of specificity, then it will not do to speak simply of valuing one's life. As it stands, that phrase is multiply ambiguous: there is, for instance, valuing *continuing* to live, valuing one's living on *as* one now does (say, at a certain level of well-being), and valuing one's living a *long time*. In different contexts, one or another such valuation may be intended by someone speaking of one's valuing one's life. It might seem that most people at least have a composite valuation of their lives formed from some process of taking account of such specific valuations. But the composition process would be complex indeed, and I doubt that life usually presents occasions for going through it. For these and other reasons, I doubt that people normally have such composite valuations.

Moreover, even how much one values continuing one's life may be partly a matter of the degree of one's aversion to what one sees as the alternative. One commonly values one thing as opposed to another, not in abstraction from everything else. If the alternative is, for instance, being killed, then one's valuation of one's life may be determined in part, negatively, by one's aversion to being killed (or, broadly speaking, one's disvaluing of being killed). To see how much difference just what is valued (or is the object of aversion) makes, imagine the difference between what you might pay to reduce the risk of dying prematurely of a sudden fatal heart attack and what you might pay to reduce the risk of dying, equally prematurely, at the hands of a brutal murderer. To speak without qualification of how much someone values his life is thus to speak at best imprecisely. One can value, or even buy, a reduced risk. One can value, and can perhaps buy, an extension of life beyond what it would otherwise be. But what would it mean to value (or to buy) life simpliciter?

In order to specify just what a person's valuations are, we must use terminology that is discriminating enough to reveal the content of the person's propositional attitude. This does not imply that people are themselves generally precise in what they value. But there are certainly many cases in which even a phrase like "continuing to live to old age in comfort and good health" is not sufficient to specify what it is that a person values in virtue of which he may truly be said to value "his life". Talk of valuing one's life, then, tends to cloak a variety of specific valuations connected with one's sense of what is worthwhile in one's life, such as valuing a kind of activity one enjoys and one's continuing to fulfill a professional or familial role.

Equally relevant to the issue here is the point that people making important decisions rarely do so in the context of valuing simply one outcome of each alternative choice, even if one outcome is dominant in their thinking at the time of decision. In buying life insurance, for instance, most people probably think mainly of the aversive consequences of dying a "natural" or accidental death without it, and give little or no thought to the possibility of being either murdered or dying, for instance, of typhoid. Similarly, in taking steps to lower their blood pressure they may not be thinking of cancer. But their negative valuations of these largely ignored aversive things, as well as of dying naturally or accidentally without life insurance or of heart disease, must be taken into account, either in order to ascertain how much they value avoiding death simpliciter, or to determine, by drawing certain inferences, how high they "price" their lives. The valuational composition process turns out to be quite complicated.

To see this last point, consider a specific case in which S's main value underlying a choice is positive, say a valuation of continued life beyond sixty-five without heart trouble (as opposed to dying of heart disease at that age). Is it possible to get a reading of how high he prices "his life" from what he is willing to pay to reduce cardiac risks to his life by a certain expected percentage, for instance by purchasing equipment designed to strengthen the heart and thereby improve his chances (he thinks) by ten percent? It is doubtful that this is possible, because the sort of general value we are trying to ascertain depends significantly on other, more specific values he rationally holds, such as his aversion to dying of cancer or perhaps to leaving his spouse a widow. Perhaps S *would* regard a risk of cancer as worse than one of heart trouble, but simply thinks he is more likely to get heart disease and so takes steps only to reduce his risk of the former. To construct a monetary valuation of his life as a whole from his point of view, based on what he pays to reduce his risk of heart trouble when he would have paid much more to reduce the *same* risk of cancer, is apparently to price "his life" too low, even from his point of view. The fact that he did not think his risk of cancer was high does not imply that one can ignore his hypothetical (presumably higher) valuation on continuing life without cancer. Again, *what really* is the relevant object of his valuation? Is it a positive state of affairs, such as living beyond sixty-five without heart trouble or living beyond sixty-five in a healthy state, or is it instead a negative state of affairs such as avoiding dying of cancer at or near age sixty-five? And how is one to tell?

It is possible, of course, to construct a questionnaire with many searching hypothetical questions. But even if, for a particular person, the results are consistent both for a given battery of questions *and* on repeated uses of the questionnaire, how should these different results be weighted? Expected variation in value assignments with increases in felt risk have already been mentioned as a problem. One might arbitrarily keep projected risks moderate or low to minimize this factor. One might also produce a composite valuation of life in the face of S's finding acceptable *different* payments for the same reduction in probability of death depending on the imagined affliction. Of course, we need not price life to get some results useful in health care planning. We could simply construct monetary valuations on avoiding, say, cancer. There are, however, many kinds of cancer and many treatments. Hence, the same problem will arise as S envisages one or another form of the disease or one or another level of care.

Nor can one neglect differences in the number, accuracy, and rationality of S's relevant beliefs, including those expressing the crucial subjective probabilities. Again, I am not claiming that rational choice requires that such beliefs meet minimal epistemic standards. My suggestion is that, morally, one should not take too seriously S's valuation of his continued life if the valuation is defective in certain ways: for instance, when it is based on misconceptions about probability, irrational probability beliefs, or foolish acceptance of views found in unreliable popular magazines. Suppose that, for foolish reasons or from insufficient attention, S thinks that a chance of one in 1,000 of dying from a health hazard is in effect negligible, and largely for *that* reason pays just $2 to eliminate it. Should one make any policy decisions regarding S, such as declaring that this risk is acceptable from his point of view or presuming that he would be willing to incur an additional one in 1,000 chance of death to save $2.50?

Note that what holds *from S's point of view* on the matter is not simply a question of what *he holds* on it. It is roughly a matter of what is indicated by a proper interpretation of his values, desires, and beliefs. This point is particularly important from a moral perspective, because taking the moral point of view surely requires that one try not to do things that a person would justifiably disapprove of if he were sufficiently clearheaded and reasonable regarding his own values, desires, and beliefs. The central moral issue is not whether he would or would not accept a given decision. That he would in fact not complain of a decision one might make is a different matter, and it sets a lower standard.

IV. A CIRCUMSCRIBED ROLE FOR MONETARY VALUATIONS IN MEDICAL DECISIONS

From what has been said, it might seem that I regard monetary considerations as either irrelevant in medical decision making or impossible to establish with any precision, or both. This is not my view. I do believe that there is so much difficulty about what is being measured, and how it is to be measured, that biases may be insufficiently restrained even in quite conscientious efforts to keep monetary considerations in proper perspective. On this liability of cost-benefit analysis, Peter Wenz has made a number of valuable points.[6] I also believe that there are special obligations in the physician-patient relationship which in general vastly outweigh any monetary considerations that normally enter into the decisions the physician and patient must make. As George Agich shows, however, it does not follow that monetary considerations have *no* moral weight in medical decision making.[7]

In exploring the moral relevance of monetary considerations, it is crucial to distinguish the level of individual decision (or small group decision) about a single patient from decisions of social policy. In the former case, there are normally special obligations such as those of physicians toward their patients. There may be implied promises; the physician will typically have laid hands on the patient; the patient will have developed trust in, even dependence on, the physician; and so on. A natural view here is that the physician should perform any service he can for the patient if the patient is competent and clearheadedly wants it. If machines are available and the patient can pay for them (perhaps through insurance), why should there be any monetary constraints? The question is not whether there should be a social policy that sets constraints; that issue will be addressed shortly. Here the question presupposes a social policy that permits using the machines in question.

Even in an individual case such as this, both the physician and the patient, as moral agents, might want to think not only about non-monetary matters such as who *else* might need the machine, but about what might be done with the money saved if the patient does not use it — and with the profit made if he does. It will, of course, matter greatly whether the machine is needed to save the patient's life or simply, as in some uses of a body scanner, to aid in diagnosis. In either case, however, the monetary implications are not morally negligible, particularly given the medical scarcity that prevails in many places where the issue could arise (for example, in large cities with substantial poor populations).

It is one thing, however, to say that monetary considerations are relevant and another to say either that they should be constructed on the basis of cost-benefit analysis or should ever be constituted by valuations of "life", say, of continued life at a specified level. Imagine that two accident victims arrive at a small hospital at once and that there is only one physician capable of tending them. Would the fact that one of them monetarily values his life more highly than the other be significant in choosing between them? Would it even be relevant? Perhaps if it were clear that each was fully rational and was also adequately informed in making the decisions on which the valuation is based, one could suppose that their different valuings might be relevant. But it may be plausibly claimed that we could not suppose them to be relevant unless the decisions they are based on satisfied *moral* criteria. If this claim is correct, then presumably the different valuings are significant because they are felt to be *indications* of the morally grounded weightings we ourselves should assign.

To be sure, the rationality conditions just specified might perhaps imply that decisions which satisfy them fulfill all of the relevant moral criteria, but I shall not assume that. However, even if only criteria of rationality need be satisfied, surely these criteria would have to be stronger than those given by maximization of expected utility theory; and if that is so, then it again appears that monetary valuation may be morally relevant only *derivatively* from the moral relevance of the different rational valuations of the two lives. Of course, if Menzel is right in asking (apparently rhetorically) what *ultimate* preferences are *not* nonrational,[8] then I am quite unsure whether the rationality of decisions underlying the monetary valuations would make those valuations relevant to a moral choice of the kind imagined between the two accident victims. (If a preference for consistency in one's outlook could be ultimate, as it apparently might be, then at least on one count, Menzel's apparent view is too strong. But he probably has in mind substantive ultimate preferences, say, for a kind of life, and there his suggestion is at least arguable.)

If monetary considerations are morally relevant at all in the kinds of individual medical decisions imagined above, then even if they do not make a moral difference to a decision by a physician and patient once treatment has begun, they might make a difference to the entire conception of treatment with which it is begun. It would be possible to become much better informed about where money for treatment comes from (for instance, how much is tax money and how much is from insurance), and where money — and time — saved by eliminating or abbreviating treatment would go. People in the developed nations do not ordinarily think of the sick and the poor elsewhere on earth (or even in their own countries) as competing for medical care. It would be quite possible, however, to turn Menzel's proposal outward and talk about the monetary value of, for instance, saving lives, by calculating how many lives would be saved by sending so many dollars worth of medicine and services to a country beset by famine and disease.

What this last set of considerations brings out is that if one takes a sufficient number of factors into account, then one can see that in a world of scarcity individual decisions reverberate far away. Such reverberations are more obvious for social policy, and I now want to consider how cost-benefit analysis might be useful there, particularly in devising a system of national health care.

V. SOME POLICY IMPLICATIONS

I first note some importantly different kinds of cases. In one kind, the task is to devise a policy for distribution of a health benefit to which no one has a right, or at least no one has a better claim than anyone else. In these cases, one might try not only for the least expensive distribution, but if a choice must be made between serving X and serving Y, then serve the one whose aid is less costly, as where an organ is given to someone nearer rather than farther from the source when all else is equal. Other things might rarely be equal, of course, and non-monetary factors by and large might be more important. But there is still scope for monetary considerations even within those constraints.

A more difficult case concerns policy regarding defective newborns, or terminally ill incompetent patients, who are wards of the state. Should monetary considerations matter here so far as resources are actually available within the hospital or institution in question? In the former case, one clearly cannot use monetary projections based on the patient's apparently rational choices, since newborns will have made no such choices. In the latter case, there is some question how the choices should be treated even if we do know them, but presumably there is a *prima facie* obligation to observe the patient's previous preferences, say, in abstaining from prolonging life.

Once again, I suspect that it is chiefly where other things are approximately equal that monetary considerations become decisive. But perhaps it is not crass materialism to suggest that they may be important even before that, if only because of scarcity. If excessive medical time, transfusions, and medication are given to one patient, others may die, or become at least as badly off as a result. The fact that one does not know beforehand precisely what the effects will be does not make generally expectable effects irrelevant. Indeed, in the case of wards of the state, while there is a special responsibility to care for them because they *are* helpless wards, the rest of the citizenry, or at least of the state's wards, may also have a special claim on the state's resources since it presumably exists for all its people. In this respect, the state's obligations to its wards are weaker than those of parents toward their dependent children.

Throughout, I have been assuming that money has no intrinsic value and that our reasons for taking it seriously concern what it can buy. On that assumption, we should be specially interested in whose money is in question and what might be done with it. There are times when medical costs exceed a patient's insurance coverage. There are also times when medical expenditures prolong a condition whose extension is neither the right nor the desire of the patient, and indeed simultaneously constitute a detriment to family or friends who must bear a difficult burden. These are all cases in which saving money can be both a morally relevant aim and medically reasonable. Many would not want certain expensive medical procedures if those procedures would jeopardize their children's education. Most reflective people do not want to have their lives extended artificially if it means becoming a burden to their families, monetarily or otherwise. If our policy were that money one saved by not doing expensive treatments could go for certain purposes, say, for extending basic medical care to those who would not otherwise receive it, then it might be more reasonable for both patients and physicians to try harder to avoid expenses whenever possible.

If what has been said about monetary considerations in relation to social policy is correct, it has yet to be shown how cost-benefit analysis bears on such policy. One point should be plain: once the relevance of monetary considerations is established by morally acceptable arguments, some form of cost-benefit analysis may be used to determine the best distribution of funds. Thus, one might: (a) consider the probabilities of getting certain results with a given amount of money; (b) value those results in relation to some reasonable conception of health; and (c) choose an allotment that is at least as good as any alternative. This procedure might lead to putting more money into preventive medicine and less into cancer research. Notice, however, that this procedure does not use cost-benefit analysis *grounded in* people's implicit pricing of their lives. Rather, mainly on *moral grounds*, it supplies valuations and uses cost-benefit analysis to maximize expected value so conceived. Is there, then, no role for people's apparent pricings of their lives in determination of health care?

Perhaps there is a significant if restricted role. Suppose that, through people's purchases of insurance, their investments in devices (like exercisers) meant to reduce health risks, and their answers to questionnaires, one could draw reasonable inferences about their pricing of their continued living. Suppose further that a huge majority implicitly placed a higher price on living out one's life to a given age without cancer than to living it out without heart trouble. Would this not provide *prima facie* reason to prefer a health care system that combats cancer more than heart disease over one with the converse emphasis? I believe it is arguable that it would. Other reasons might take priority, but this would be a significant one.

On the other hand, I cannot see that this method would imply anything significant about just *how much* to put into either domain. Governments could perhaps view the matter this way: calculate what percent of people's spendable resources they voluntarily use to reduce certain risks (taking account of both their evaluation of the relevant aversive outcomes and of their assignments of probabilities to those outcomes), and then devote a proportionate share of public expenditures to health care policies combatting those risks. Such a policy, however, would be at best conjectural. Many people might prefer, and perhaps rationally prefer, that society as a whole spend a greater or less proportion of its funds on cancer prevention than they do themselves. In any event, should not citizens or at least their elected representatives be asked such questions?

Surely they should, and this brings me to my final point regarding public policy. If monetary valuations of "life" can be reasonably extrapolated, would it be *better* policy to use them as a guide in shaping the health care system than to elicit public debate and try to make rational decisions on the basis of a vote, even if only of elected representatives? I suspect that the latter procedure would be preferable, and it at least comes closer to getting people's informed consent to health care decisions affecting them than does basing the relevant decisions on their monetary valuations of their lives, or of their health, extrapolated from their patterns of expenditure and responses to questionnaires. Granted, there might still be a role, where other things are approximately equal, for a comparative funding decision to be influenced by the monetary valuations people place on the competing services under discussion. If, for example, people spend far more to reduce cancer risks than risks of coronary artery disease, that would, other things being equal, favor larger amounts of research expenditure in the former domain than the latter. It is an empirical question just how important that role would be in actual cases. It is by no means clear that it would be major in a society such as that of the United States today.

VI. CONCLUSION

I began by locating the decision-theoretical roots of cost-benefit analysis as a strategy for rational choice. I noted a number of difficulties with the basic model, particularly from the moral point of view: first, its apparently narrow application, given the paucity of probability beliefs on the part of most people with respect to at least some significant possible outcomes of the options they consider in at least their typical decisions; second, its taking no account of the rationality of the beliefs and valuations constituting the basis of the crucial calculations (or even the truth of those beliefs); third, its failure to distinguish between a choice's being *rational for* a person from that person's *rationally making* that choice; and fourth, its inability — even if it should provide a good model of one kind of rationality — to provide sufficient conditions for choices of a kind that yield a *morally* acceptable basis for medical decisions concerning the subject. Regarding this last limitation, I stressed the analogous point that even informed consent that is rational from one or another decision-theoretic point of view may not be warranting consent, that is, the sort that morally validates using a medical procedure to which the subject has so consented. I also stressed the difficulty of characterizing tacit consent, and I argued that people do not tacitly consent even to everything clearly implied by what they explicitly consent to. They certainly do not automatically consent to accepting, for a certain monetary saving, risks they have apparently priced at or below the relevant monetary value.

On the other hand, not every use of monetary considerations in medical decision making must be regarded as objectionable. While money is not intrinsically valuable, it is a means to things that are, including restoration of health and happiness to people who could not be treated unless someone else's not being treated — possibly in order to save money — freed the needed resources. It appears, however, that cost-benefit analysis not only fails to provide the criteria for deciding when and how monetary considerations are relevant to medical decisions, but also has a limited role in determining how money should be spent even where more of it may justifiably be used for one important medical purpose than for another. Cost-benefit analysis may help us in ascertaining efficient uses of funds, but not their ultimate destination.

Decisions regarding where funds are needed in the health care system should not depend on people's pricing of various elements in the health care domain. One problem is the difficulty of ascertaining this pricing at all from the limited information available, given people's probably frequent deviations from the cognitive behavior one would expect if they were in fact maximizing expected utility. But even where one can determine their pricing of one or another aspect of their lives, its significance for medical decisions is limited in at least the ways indicated above.

To be sure, in a few instances cost-benefit analysis may yield findings quite important for medical decisions, and may even do this through projections of people's monetary valuations of their own continued living. This point seems to hold particularly where a huge majority of the population can be plausibly seen as monetarily valuing one element in health care over another. Other things equal, a comparative decision favoring the former domain over the latter might be warranted, even if there remains a problem of determining the specific allotments to each medical need. But, far more often, policy decisions must be made on the basis of much less nearly quantitative data, and they involve weighing a variety of competing considerations. Containing costs is an important goal here; and cost-benefit analysis can be a valuable corrective to the high-minded view that, almost without exception, patients must be given every service possible. The best course seems to be to encourage public discussion, to bring together philosophers, physicians, and others to study the problems, and to develop a framework for a making medical decisions, individual and social, on a carefully weighed complex of moral criteria.[9]

University of Nebraska
Lincoln, Nebraska

NOTES

[1] This is argued in detail, and in relation to epistemic rationality, in [2].

[2] For critical discussion of these problems besetting the maximization of expected utility conception of rationality, and for a sketch of an alternative account of rationality, see [1].

[3] See Paul Menzel's essay, 'Prior Consent and Valuing Life', note 9. See also [4], pp. 30–35.

[4] This seems to be Menzel's strategy in ([4], p. 22).

[5] It will be clear later that I do not agree that the only alternative to taking this view is doing little or nothing to contain health care costs.

[6] See Peter Wenz' essay, 'CBA, Utilitarianism, and Reliance Upon Intuitions'. There is much in this paper that I accept, and I especially appreciate Wenz's emphasis on treating some of S's preferences more seriously, in moral decisions about S, than others. I would deal differently, however, with some of the issues and would distinguish explicitly betwen reliance on *intuition* and reliance on *conjecture*, though the room left for prejudice and arbitrariness may often be approximately equal in both cases.

[7] See George J. Agich's essay, 'Economic Cost and Moral Value'. The attempt to show that even a Kantian view does not preclude giving monetary considerations some moral weight is especially noteworthy. Edmund Pellegrino has offered a different perspective,

stressing the primacy of moral and other non-economic values in medicine; for a short statement, see [5].

[8] In his essay Menzel also says that he is not committed to ignoring irrationality or to giving different weights to irrational preferences as opposed to rational or non-rational ones. Richard Brandt develops an account of how we may determine what ultimate preferences *are* rational, and he treats non-rational elements as equivalent to irrational ones [3]. In [1], I assess Brandt's approach and suggest some alternative ways of making sense of rationality for intrinsic, hence in a way ultimate, values and desires. Here I differ considerably from Menzel, but I emphatically agree with him in distinguishing non-rational from irrational elements.

[9] For helpful discussion, I want to thank George Agich and Heidi Malm.

BIBLIOGRAPHY

[1] Audi, R.: 1985, 'Rationality and Valuation', in G. Seebass and R. Tuomela (eds.), *Social Action*, D. Reidel Publishing Company, Dordrecht and Boston, pp. 243–277.

[2] Audi, R.: 1985, 'Rationalization and Rationality', *Synthese 65*, 159–184.

[3] Brandt, R.: 1979, *A Theory of the Good and the Right*, Oxford University Press, Oxford.

[4] Menzel, P.: 1983, *Medical Costs, Moral Choices*, Yale University Press, New Haven and London.

[5] Pellegrino, E.: 1978, 'Medical Morality and Medical Economics', *Hastings Center Report 9*, 8–11.

PART III

ECONOMICS AND ETHICS IN HEALTH POLICY

J. MICHAEL SWINT AND MICHAEL M. KABACK

INTERVENTION AGAINST GENETIC DISEASE: ECONOMIC AND ETHICAL CONSIDERATIONS*

I. INTRODUCTION

In this paper we examine some of the economic and ethical consequences associated with genetic screening and intervention — a field that is expanding rapidly as technology is making possible the prenatal detection of an increasing number of genetic diseases. In 1983 the March of Dimes Birth Defects Foundation conducted a worldwide survey which revealed that 352 medical genetics facilities in 45 countries provide prenatal diagnostic services in early pregnancy [22]. These developments have all occurred since the first reports of intrauterine fetal kerotyping in 1967. We provide information on the trade-offs between economic and ethical consequences of expanding the screening efforts for three genetic diseases. We believe that these types of considerations are particularly important given the rapid advancement and diffusion of technological capabilities in this field. Victor Fuchs correctly asserts:

> No nation is wealthy enough to supply all the care that is technically feasible and desirable; that is, to supply presidential medicine for all. However, it would be a great mistake to turn physicians into explicit maximizers of the social cost benefit ratio in his or her daily practice. But the trade-offs must be made. Usually the best time for making such decisions is during the evaluation of the costs and benefits of new facilities and the development and diffusion of new technologies ([10], p. 1572).

In recent years our society has enjoyed significant increases in the availability of health care services and yet often has been able to avoid explicit consideration of difficult economics/ethics health policy trade-offs. That is, to a large extent we have had the freedom to ignore the costs of health care for insured segments of the population.[1] It is becoming increasingly difficult, however, to retain this situation with the electorate and various political pressure groups calling for a decrease in the growth rate of health care expenditures. This is resulting in additional attention being given to the economic consequences of utilization decisions. If consideration is given to economic consequences, then to

G. J. Agich and C. E. Begley (eds.), The Price of Health, 135–156.

the extent that indicated policy directions are not coincident with indications given by ethical and medical concerns, conflicts and trade-offs will arise.

In essence these trade-offs are concerned with the determination of society's prices of health — how much are we as a society willing to spend on health care services? Lester Thurow asks: "As a society, how much are we willing to spend (that is, sacrifice) to prolong life? The easy answer is any amount, but that answer is neither true nor feasible. Like it or not, Americans are going to have to come to some social consensus concerning the trade-off between costs of medical services and the life-extending benefits that result" ([40], p. 1571). Indeed, expenditures for health care are currently approaching 12 percent of GNP and already systemic cost containment policies have been enacted and others are under consideration. There is certainly no intrinsic limit to the percentage of GNP that should go to health care, but certainly we will reach levels of health care expenditures beyond which society will demand some restraint. How close we are to that limit is something that society will have to decide; in doing so it will effectively be helping to determine its "price" of health. This is something that we have typically avoided facing in the past, in most instances we have even avoided discussing it.

The consequence of determining the price of health which sets limits to resource usage in the face of medically indicated need is, of course, rationing. This, too, is a typically avoided subject, for example, witness the President's Commission avoidance of the term rationing and its primary explicit emphasis (in the economic sphere) on the elimination of inefficiencies.[2] While the elimination of inefficiencies is important, in the long run it is unlikely to be enough.[3]

The pressures to lower the growth rate of health care expenditures, in combination with the growing availability of efficacious medical technologies and the aging of the population, are going to make avoidance of this issue and explicit confrontation with the rationing of medical care increasingly difficult. There are also costs associated with avoiding the rationing issue. Mary Ann Baily puts the problem succinctly: "As a result of our ducking any implication of rationing, the limits are imposed in less visible ways, and they are imposed most heavily on those in society who are least able to resist" ([4], p. 498). If society is to determine openly the price to be placed on health, it will have to engage in serious discussion and debate. Because much of this discussion will involve empirical as well as philosophical questions, information on the assorted consequences of alternative courses of actions will be required. These trade-offs remain relatively unexplored and deserve increased attention.

Here we do not offer a critique or a justification of the use of economic and/or ethical consequences as social objectives in planning interventions against genetic disease. Rather, we attempt to delineate some of the areas of conflict and congruency among these considerations as they relate to policy alternatives for the allocation of resources in the field of genetic screening. The President's Commission for the Study of Ethical Problems in Medicine and Biomedical and Behavioral Research noted

> ... the fundamental value of genetic screening and counseling lies in its potential for providing individuals with information they consider beneficial for autonomous decision-making. Therefore, although the societal impact of genetic screening and intervention and cost-effectiveness are indeed relevant considerations, the benefits and harms that accrue to the individual screenees deserve special consideration ([33], p. 97).

Indeed they do. But how far do we go in the provision of services under varying ethical and economic circumstances? It is a social decision as to (1) whether we will sacrifice achievement of certain ethical concerns for economic gain or (2) how much economic loss society will tolerate to achieve its explicit ethical objectives. It should be determined by a society that is aware of the consequences of its decisions.

II. ETHICS AND ECONOMIC TRADE-OFFS

Efforts to develop and promote prenatal evaluative technologies are motivated by the desire to alleviate personal and social problems. In genetic screening and counseling, providers are concerned with (1) protecting autonomous decision making, (2) preventing harm, (3) equitable provision of screening and counseling services, and (4) maintaining confidentiality of test results. We will explore the extent to which these ethical concerns are in conflict or are congruent with policy indications given by economic criteria. One objective of this type of exploration is the eventual identification of policies that do not yield such conflict and thus the recommendation to increase funding and/or third party insurance coverage where indicated. If we can pursue policies that result in increased economic efficiency without violating ethical concerns, then a major concern with relying on economic criteria would be removed. If we find instances in which ethical concerns are seriously violated, we should attempt to find ways to correct this situation — for example, by redesigning the policy. If this is not immediately possible, then a social policy trade-off must be faced and further debated.

There are currently several thousand identified "genetic diseases", and many more will be identified each year. Some of these diseases have a uniform prognosis, while others are quite variable in their expression. The economic and ethical consequences vary across the disease spectrum. As space limitations preclude examining them all, we focus on three genetic diseases in order to discuss and illustrate some of the issues involved. The first disease, Tay-Sachs, while presenting some serious ethical issues, is relatively straightforward in terms of economic ethics policy convergence (and thus it deserves more favorable government and insurance treatment than it has thus far received). The second disease, Down's Syndrome, presents somewhat more difficult trade-offs. The third disease category, Neural Tube Defects, presents some very difficult economics/ethics trade-offs with which society must wrestle.

Much of the controversy surrounding the intervention against genetic disease involves very real and justifiable differences of opinion as to the degree of disability associated with a predictable pregnancy outcome (for example, Down's Syndrome detected prenatally) and/or when the results of a prenatal examination indicate the presence of disease in an affected fetus, but not the severity that can be expected from the disease's range of possible outcomes (as in Neural Tube Defects). The diseases chosen allow examination of this spectrum of possible outcomes.

III. ECONOMIC CONSIDERATIONS IN GENETIC SCREENING AND COUNSELING

The economic evaluation of health care policy alternatives involves a comparison of the economic benefits derived from interventions with the benefits society must forego by not using these resources for alternative purposes (opportunity costs). Cost-benefit analysis (CBA) is utilized to achieve this comparison. The application of CBA to prenatal diagnosis of genetic disease does *not* in itself imply that economic consequences form the decision-criterion for program approval.[4] Rather, CBA is a decision-assisting technique that provides decision makers with information designed to improve their ability to make rational decisions. In other words, CBA functions only as one source of information that decision makers must combine with ethical, socio-political and other information for net evaluation. This point is clearly stated by A. Prest and R. Turvey in their classic evaluation of the utility of CBA: "It is only a technique for making decisions within a framework which has to be decided upon in advance and which involves a wide range of considerations, many of them of a political or social character" ([34], p. 733).

Thus, it is important to maintain the distinction between public policy decision makers and the analysts who provide them with various types of information. Much criticism of CBA in the health field has come from those who have (perhaps inadvertently) set up a "straw horse" by incorrectly asserting that CBA is a decision-making technique, and then, having postulated this, proceeding to criticize its applications as inappropriate. It should be noted that whenever program evaluation decisions are made without the benefit of net economic impact data, an implicit economic value has been assumed nevertheless.

The aggregate economic impact of genetic disease is substantial. Based on a study of 43,558 infants, the overall frequency of chromosomal abnormalities is about 5.6 per 1,000 births in the United States ([21], p. 4). The frequency of major congenital anomalies is about three percent among newborn infants. While these figures are evidence of relatively rare outcomes, the aggregate impact is not negligible; that is, an estimated 20,800 infants in 1975 were born with chromosomal abnormalities in the United States ([8], [24]).

The net future commitment of society to the maintenance and care of individuals affected with chromosomal abnormalities alone is in excess of $7 billion per year (adjusted to 1986 prices) [24].

Institutionalization and/or special education must also be considered. If an infant affected with a genetic disease was institutionalized in 1972 for a period of seventy years, the discounted present value of this cost (in 1972 prices) was $149,000 [37]. Since most do not survive this long and would spend a significant amount of time away from an institution, the cost is overstated. If we assume that an individual with Down's Syndrome spends an average of twenty years in an institution, between the ages of ten and thirty, the present value of the cost of care would be $65,000.

Finally, there are medical care costs incurred for the care of affected individuals. In a study of the costs of pediatric care for infants with genetic disorders, J. Hall *et al.* found that these patients have many more admissions which are slightly more expensive; that they stay longer in hospital; that they travel farther; and that their families end up paying the bill more often [12]. However, the variation in medical care costs between diseases and often between cases of a particular disease is great. They range from about $2,200 (exclusive of physician fees and clinic costs) per hospitalization for sickle cell patients (1974 figures adjusted to 1986 prices) [39], to $20,000 per year for hemophilia [26], to as high as $60,300 to $120,600 (discounted present value) for just two and one-half years of care for a case of Tay-Sachs Disease (1976) [28].

Thus, with respect to prenatal intervention in genetic disease, while the decision to undertake intervention is highly individualistic, there remain collective concerns since such decisions have consequences beyond those borne by the couples in question. Even if a couple can afford to pay the monetary costs of caring for an affected child, the medical (and other) resources involved are precluded from alternative uses. Thus, from the social perspective, the measurable economic benefits of diagnosis and (if indicated) intervention include reduction of medical care costs, avoidance of costs of institutionalization, and special education and training. In addition, there is avoidance of risks to mother such as toxemia associated with carrying a fetus affected with Barts hydrops fetalis, and avoidance of interruption of normal pregnancies (as with pregnancies that previously may have been interrupted due to the high risk characteristics of the mother and the prior absence of prenatal diagnostic techniques). The economic costs of intervention include program (or direct) costs: counseling, diagnostic procedures, intervention procedures; indirect costs to patients: time loss, travel, etc.; misdiagnoses of procedures and their economic consequences; and, risks of procedures to the mother and to the fetus and their economic consequences.

In general, for a genetic disease to be suitable for prenatal diagnosis and intervention from the economic perspective, it should be confined to a well-defined population,[5] lack a curative treatment for affected infants (except where prenatal diagnosis increases the probability of successful postnatal treatment), and represent a substantial economic burden to society. While there are many genetic diseases that meet these criteria, intervention is presently very limited, given what is technically possible. As such, evidence of its economic impact, while generally favorable, is limited.

IV. TAY-SACHS DISEASE

Until recently, Tay-Sachs Disease (TSD) was the only single gene disorder detectable by amniocentesis for which there have been widespread carrier detection efforts. The disease is a fatal autosomal recessive disorder that affects approximately one of every 3,600 infants born of Ashkenazic European (Ashkenazic Jewish) descent. Affected infants typically die before the age of four, after enduring a lengthy period of suffering. In addition to the emotional suffering the family must bear, there is also a considerable financial burden due to the repeated hospitalizations that are required during this period. Third party insurance coverage is clearly inadequate. As TSD occurs predominantly in this defined population group, healthy adult carriers of the disease gene can be readily and accurately identified, and since the disease can be detected in the affected fetus in early pregnancy, the condition becomes well suited for mass carrier screening among Jewish adults. In the past decade, in the United States, nearly 500,000 child-bearing-age Jewish adults have been screened voluntarily to determine their carrier status. Cost-benefit analyses have shown that these programs, in combination with diagnostic and therapeutic abortion services for couples at risk, have positive economic consequences. Evaluations of two separate programs found benefit-cost ratios ranging from 3:1 to 10:1 for community screening efforts (with commmunity organization and publicity) and from 1.6:1 to 3.2:1 for on-demand hospital screening ([28], [38]).

Additionally, O'Brien has estimated the economic impact of screening for TSD on a nationwide basis [30]. For prenatal diagnosis and pregnancy interruption if indicated in families with a previous TSD child (18 percent of cases are in such families), the benefit-cost ratio was 54:1; for heterozygote detection in relatives of affected persons who are $\geqq$ 30 years, with amniocentesis and interruption where indicated, the benefit-cost ratio was 13:1; and for screening and intervening for all U.S. Jews < 30, the benefit-cost ratio was 11:1. Thus, while the number of TSD cases per year is small, the risk group is well enough defined for a screening program to have positive economic consequences. As more screening occurs, families in which the trait occurs are identified, the carrier detection rate (per number screened) increases, and costs per TSD case detected decline. The current indication therefore is to make carrier screening (and amniocentesis) and counseling services available to Ashkenazic Jews of child-bearing age. As educational efforts continue, it is anticipated that the demand for TSD screening will grow and that similar services will become more available.

While considerable attention has been given to the possible social stigmatization of individuals found to carry such recessive traits, research which addresses these issues (among program identified TSD carriers) has failed to identify any clear evidence of such a phenomenon ([23], p. 117). It has been suggested that effective pretesting education and post-screening counseling serve to obviate or minimize this theoretical concern. However, there may be an important "temporal" component in possible stigmatization of carrier-identified individuals. That is, individuals may not feel the "burden" of being a carrier until the time arises when they confront either marital choices or reproductive decisions. In this context, long-term follow-up studies of identified carriers have not been conducted as yet nor have conclusive data been presented to ensure that such phenomena do not occur.

A second issue with regard to carrier identification, stigmatization, and other attendent ethical issues relates to the use of centralized (name-associated) data banks or registries. Such registries probably should be avoided since they are likely to provide access to carrier-related information about individuals. This could create an important breech of confidentiality and/or privacy for those individuals tested in large-scale screening programs. Rather, it may be appropriate for statistical purposes that only coded or non-identifiable aggregate data be included in such registries.

Another concern is that some people may not understand what they are told by the screeners. They may become confused and worried and, therefore, may not make effective and rational decisions. Carefully planned counseling efforts are needed to minimize this possibility.

Clearly, the avoidance of TSD involves the reduction of mental pain and suffering of the parents who elect to have abortion performed and thus avoid undergoing two to three years of life with a suffering and dying infant. The same argument can be made for the infant — avoidance of the two to three year period of suffering is the appropriate alternative. There is, of course, a very small, known risk involved in the amniocentesis procedure itself. This is a risk of which the couple has been informed and has incorporated in their decision to utilize the procedure.

With respect to privacy, careful controls can help assure the confidentiality of all results; however, the psychological stigma may be real. This can be partially overcome through counseling efforts but it also must be balanced against the potential medical and humanitarian value associated with the acquisition of this information. The expansion of TSD screening efforts clearly has improved access to these services and thus has made the distribution of the services more equitable.

As we suggested earlier, we believe the TSD case is relatively straightforward. With only minor and surmountable difficulties, TSD screening programs result both in positive economic and ethical consequences.

V. DOWN'S SYNDROME

Only 22 percent of Down's Syndrome births in the United States are to women over thirty-four years of age, a reflection of the fact that the proportion of total births to women in this age category has declined by more than fifty percent since the 1960s. Thus, the potential impact of programs focused on this risk group alone is less than one-quarter of the annual incidence of the disease [1]. However, each year, in the United States, another 700 to 800 cases of serious chromosomal abnormalities could be detected with the same test. Down's Syndrome and other chromosomal trisomies do meet economic criteria for intervention and there is evidence to indicate the economic desirability of expanded efforts in this direction.

In a study of 526 cases of amniocentesis with detection of 16 affected fetuses, R. Conley and A. Milunsky found benefit-cost ratios for Down's Syndrome and Trisomy 18 interventions of approximately 2:1 [8]. The general indications of these results are reinforced by a Center for Disease Control (CDC) study in which the benefit-cost ratio is 1.5:1 (for maternal age $\geqq$ 35) [6].

While the evidence to support the economic viability of amniocentesis for women ≧ 35 years is clear, it has been established by cohort analysis; for example, women aged 35 and 39 are in the same cohort risk group. There is need for a more precise measure. E. Hook and G. Chambers estimated the incidence of live born Down's Syndrome infants by yearly maternal age, for mothers aged 20 to 49 [14]. Evidence from E. Hook and J. Fabia's analysis indicate that maternal age as low as 33 or 34 years may be appropriate for amniocentesis from the economic perspective [15]. At a minimum, this supports previous findings of the economic viability of amniocentesis for women ≧ 35 years and also lends sufficient support to the inclusion of women aged 33 and 34 such that the issue deserves further examination. Interestingly, Hook and Chamber's ". . . graphed reported rates (Down's Syndrome by maternal age) suggested a linear increase in rates between 20 and 30 (or just over 30) and a logarithmic increase in rates from 33 to 49 with a transitional region between these intervals" ([14], p. 127).

Consideration should be given to the potential danger to the parents where the possible birth of such an affected fetus indirectly leads to the judgment that having a Down's Syndrome child is to be avoided at all costs; at the very least, this would impair the autonomy of their decision making. This point goes beyond insuring educational objectivity in private discussions with counselors and includes concern over "media hype" about the issue. In the desire to make prenatal screening services available, there is the implicit message that a Down's Syndrome child is not worth having. It may not be possible to advocate the former without to some extent signaling the latter. This is simply an ethical problem that must be faced.

The issue of preventing harm vis-à-vis the rights of the unborn is another important concern. This, of course, is a difficult issue since the disease is not fatal and an "acceptable" life for the affected can be achieved in many instances. But the point is in conflict with the autonomy of decision making for the pregnant woman – at least those pregnant women who would choose to abort. One difficulty is that there is a continuum of sequelae among various genetic diseases. Whereas TSD is fatal early-on, another disease may have but minor consequences; still other diseases may have very serious consequences short of early death. In terms of prevention of harm to the unborn, where do we draw the line? On what criteria should it be based? By whom should it be determined? That is, when is abortion ethically justified? The literature is replete with lengthy and conflicting analyses and discussions of this issue; it is not our intention to add to them here.

As suggested by the Antenatal Diagnosis Consensus Development Task Force, there is considerable public sentiment favoring women's freedom of choice where the fetus has a serious genetic disease [41]. Since abortion of nonviable fetuses is constitutionally protected, the ethical problem usually falls on individual couples who must make the decision in accord with their own conscience. The ethical justification for aborting an affected fetus is often thought to be the moral obligation to prevent suffering. While the issue is by no means a simple one, this justification is apparently widely accepted. Furthermore, the Task Force states that the ethical issues raised by abortion merge with social, legal, and public health considerations; fundamental to all of these is the issue of a pluralistic set of values. There are important values behind the conflicting positions. As suggested by T. Powledge and J. Fletcher, the "... profound conflict of values embodied in the abortion debate ... cannot be finally resolved in a set of guidelines that offer diffections for choices that must always be made in the light of competing, and sometimes polar claims" ([31], p. 171). Their guidelines are respectful of pluralistic values and are intended to help counselors develop the favorable circumstances for thoughtful, informed, morally responsible decision making by parents. The Antenatal Diagnosis Consensus Development Task Force adopted this as the appropriate context in which to view the issue of abortion of fetuses affected with serious genetic disease and we adopt it as well for this paper [41].

In terms of equity of the access to services, the increased availability of Down's Syndrome services over the past decade clearly has increased the access of certain segments of the population. However, as the demand for these services has rapidly grown, in some instances to exceed available supply, services have had to be restricted. One consequence has been for those under 35 years of age to be discouraged from having the test (by counselors and/or physicians) or actually turned down upon request. To insure equity in the availability of services, further growth in the availability of these services seems to be required.

The cutoff point of 35 years of age for recommendation of amniocentesis initially may have been set for reasons that pertained to perceived risk of amniocentesis and probability of Down's Syndrome by maternal age (although economic considerations were apparently involved) ([33], pp. 75—81). Data now indicate that the risks (of the procedure) are lower than initially was believed. This fact and the improved data on probability of Down's Syndrome by maternal age make the maintenance of the 35-year-old cutoff more difficult to justify ([13], [14]). In addition, it is clear that the elimination of Federal funding for abortion for the poor reduces their access to services. Since these services are more readily available to those who can pay, there is a serious inequity of access to these services.

It appears that an expansion of the availability of these services would remove this ethical concern and at the same time improve economic efficiency, for while the 35 years and older criteria may have been in part retained for reasons of perceived efficiency, the studies cited above seem to indicate the contrary — an expansion of services to pregnant women 33 and 34 years (on their demand) would be economically efficient.[6]

VI. NEURAL TUBE DEFECTS

Neural tube defects (NTDs) are the result of failure of closure of the neural tube. From one to three percent of infants are born with significant malformations, of which defects of the central nervous system (CNS) and cardiovascular system are among the most serious. The neural tube defects typically include anencephaly, encephalocele, and spina bifida, the most frequently encountered of the CNS malformations.

Anencephaly is the partial or complete absence of the cranial vault with a degenerative or rudimentary brain. This defect is lethal, but 25 percent of such fetuses are born alive. It is rare for these infants to survive more than a few hours from birth; many are stillborn and others are aborted spontaneously during the pregnancy. An encephalocele is the protrusion through the skull of the brain and its covering membranes. It may be compatible with survival depending on its size, and it may be associated with other anomalies. Serious neurologic deficit including mental retardation frequently results. Spina bifida results from a failure of closure of the neural arches and most often occurs in the lower spinal region. About 85 percent of these defects are spina bifida cystica (myelomeningocele). Of infants affected with this defect, eight percent die or are stillborn and 80 percent of those surviving to age five are severely handicapped [9].

R. Althouse and N. Wald followed 102 cases of open spina bifida cystica and eight percent had no handicap; however, 64 percent died before age five [2]. Of 111 cases of closed spina bifida cystica they followed, 40 percent died before age five and of the remaining 60 percent, one-third had no handicap, one third was moderately handicapped and one-third was severely handicapped. A. Sadovnick and P. Baird found in British Columbia that 60 percent of spina bifida infants died before age seven and the survival curve reached a plateau between ages seven and thirty for the remainder, which was followed by increased morbidity and mortality due primarily to renal failure [35]. The remaining 15 percent of spina bifida defects are spina bifida occulta, a condition which rarely leads to any neurologic deficit and may be an x-ray finding only.

Spina bifida cystica varies in extent, level, and association with other malformations. Small, well-covered, low lesions without other malformations may allow normal development with little physical handicap. Others (estimated to be 90 percent) lead to seriously handicapping neurologic impairment of the lower extremities, impaired bowel and bladder function, and recurrent urinary tract infections as a result of sphincter paralysis. Hydrocephalus is a frequent complication and is seen in about 70 percent of patients with myelomeningocele. About 50 percent do not survive beyond two years of age and approximately 25 percent of the survivors are mentally retarded [7].

The true incidence of NTD is not known since nearly 50 percent of anencephalies are spontaneously aborted. In Maine the prevalence of NTDs is estimated at 3.4 per thousand, in California 1.1 per thousand, and on Long Island 1.2 per thousand [36]. In the United States there is a general decrease in prevalence as one moves from north to south and east to west. The overall prevalence in the United States is estimated to be between 1 to 2 per 1,000 newborns (3,000 to 6,000 births per year). In certain areas of Great Britain the prevalence has been estimated to range from 3.4 to 5.1 per thousand and in certain areas of Canada 3.0—4.5 per thousand.

During early fetal life, the major serum circulating protein in fetal serum is alpha fetoprotein (AFP). When there is a disruption in the fetal integument this serum protein diffuses into the amniotic fluid surrounding the fetus. In turn, AFP is absorbed through the placenta and membranes into the maternal circulation. Although it is normal for low levels of AFP to be found in both amniotic fluid and maternal blood throughout pregnancy, these AFP levels are found to be elevated highly when open lesions of the fetal brain and spinal cord are present. Accordingly, measurement of AFP levels in maternal serum has been employed as a screening test for increased risk for fetal neural tube defects.

The quantification of amniotic fluid (AF/AFP) is utilized by nearly all United States centers which perform midtrimester amniocentesis, and is now regarded as an accepted practice. In many areas, centers mail AF samples to one of the 10 to 15 laboratories with developed expertise in performing this measurement [41]. Current recommendations are that routine amniotic fluid AFP measurements of AF be conducted whenever amniocentesis is being performed for another reason (for example, maternal age). It is also recommended for pregnant women if the couple previously had an affected child.

When maternal serum AFP (MS/AFP) testing became available, it was not recommended as a routine test during pregnancy. However, a number of recent pilot studies in the United States and Canada revealed mass testing to be safe and efficient. The report of a Consensus Meeting concluded that:

> Large scale MS-AFP screening of pregnant women between 16 and 20 gestational weeks is able to detect virtually all anencephalic fetuses, a majority of fetuses with open spina bifida, about half of all cases with abdominal wall defects .... In general the sensitivity has been found to be greater than 90 percent for anencephaly and more than 70 percent for open spina bifida. Closed spina bifida ... usually does not lead to elevated MS-AFP ([36], p. 83).

Beginning in January, 1986, in California, all women who are 16–20 weeks pregnant are being offered maternal serum AFP testing. Informed consent procedures will be carefully followed and the $40 per case costs of the test will be covered by third-party payers. This fee covers any subsequent procedures necessitated by a positive test result. It thus appears that the recommendations for maternal serum AFP testing are in a transitional state and if expanded pilot studies (such as the one in California) show positive results, it is conceivable that eventually this test will be made available to all pregnant women in the United States.

If all United States pregnancies (approximately 3,000,000 per year) were monitored by maternal serum AFP (of an expected 4,500 NTDs), approximately 2,025 or 90 percent of all anencephalies and 1,800 or

80 percent of all spina bifidas could be identified. This would require an estimated 60,000 to 90,000 sonograms per year and 40,000 to 60,000 amniocenteses. (It should be noted that only five percent of infants born with neural tube defects are born to couples with a prior history of these anomalies in themselves or their children [9]). With the cutoff value currently recommended, false negatives could occur in approximately 20 percent of fetuses, while false positives are very unlikely.

R. Layde *et al.* estimated the economic consequences of a multi-tiered program utilizing maternal serum AFP screening, ultrasonography, and amniocentesis (where indicated) for a theoretical cohort of 100,000 women in the United States [18]. Assuming 100 percent utilization by those willing to participate in the program, program benefits have been projected to exceed costs by approximately 2 to 1. If these results could be generalized, it would cost over $60 million (1978 prices) to test all pregnancies in the United States in this manner, with expected economic benefit approximately twice that amount. This projection does not necessarily imply that such a policy should be undertaken, because concerns do remain, particularly with respect to procedural risks [7].

A prior study conducted in Scotland by S. Hagard *et al.* did not yield favorable economic results. It is likely, however, that this was due to the lower costs of care and institutionalization in Scotland and the fact that this study (1974 data) was done prior to the availability of clearer evidence of the sensitivity of maternal serum AFP screening.

Sadovnick and Baird concluded that maternal serum AFP screening program in British Columbia would be cost-effective. They estimated lifetime costs of care for an average case of spina bifida to be $84,000, and that for an MS/AFP screening program for NTDs benefits exceeded costs 1.8:1 [35].

The use of amniotic fluid AFP to detect Neural Tube Defects *in utero,* until recently, has been limited primarily to women with previously affected pregnancies. These women have a two to five percent recurrence rate. Fifty percent of fetuses with neural tube defects are either stillborn or die within three days of birth, 25 percent do not live beyond to age of five, and all of those that do live beyond five years have various physical ailments (of varying degrees), and one-half of those are mentally retarded. The economic costs of these events are high, and given the favorable findings of R. Layde *et al.* for a program with greater costs including maternal serum and ultrasound screening and lower incidence (it was not restricted to women with a family history for Neural Tube Defects), the net economic impact of the testing women with previously affected pregnancies for neural tube defects is clearly positive [18].

The report of a Consensus Meeting held in Germany concluded:

> ... all accounts demonstrated that the actual cost per case detected is considerably lower than the cost of medical care of a survivor during an average life-span. This holds even if one considers only detected cases of open spina bifida. The balance of costs and possible pairings depends not only on the sensitivity of the test but also on the prevalence of open spina bifida in the screened population. The favourable benefit to cost ratio is notable for countries with a high prevalence of NTD, but the balance remains positive even for a population with a birth prevalence of NTD of about 1 per 1000 ([36], p. 83).

These figures include the United States. The report further concluded that "ongoing field trials should be continued and scientifically controlled. If possible they should be extended gradually to cover the entire pregnant population" ([36], p. 83). This meeting, as well as results of studies reviewed earlier, indicate that the economic indications support current use in the United States and also prospectively support expanding the availability of services to all pregnant women.

There is serious conflict, however, from the standpoint of ethics, and we do not offer any resolution. Appropriately, this continues to be the subject of social debate. However, given the rapid expansion of NTD screening in the United States, in terms of pilot programs and the statewide program beginning in California, the weight of opinion apparently favors expanded screening. Even given this tendency, it is very important that serious efforts are taken both to assure that the screening is completely voluntary and to educate parents as thoroughly as possible to enable them to make informed, rational decisions.

The extension of screening for NTDs has provided parents with information that will allow them to make more informed, autonomous decisions regarding the pregnancy. Public education programs and community-based screening efforts have allowed increased access to services that previously were available only to well-informed couples. Access also has improved for lower income and less educated couples, and thus equity in the distribution of these services has improved. This improvement, however, is tempered by the fact that many women still do not have access to any prenatal care. Without prenatal care women will not receive counseling regarding the availability of maternal serum AFP testing. Furthermore, women with minimal formal education may, even with counseling, gain only an inadequate understanding of the complex issues involved in making an informed, autonomous decision. This presents obvious problems with respect to informed consent procedures.

For many couples, however, there will be improved access to services and improved autonomy of decision making. This advantage will have to be weighed against concerns regarding the low probability that an aborted fetus with a true positive NTD test might have had relatively minor physical and mental handicaps if carried to term. That is, there is considerable controversy over whether NTD screening and intervention should be undertaken at all, given the wide range of severity of sequelae that is possible once a fetus is found to be affected. The difficulty is that the screening tests do not identify the severity of the disease, but only its presence or absence. Since the great majority of affected infants either die before the age of five or are moderately to severely handicapped, there is a small proportion of affected infants that does not suffer from a serious permanent handicap. Thus, the difficulties with regard to the continuum of severities of possible outcomes (given a variety of different diseases) are brought together here for one particular class of genetic disease.

While the economic indications for current uses of NTD screening and for expanded community wide screening are positive, there is controversy regarding the ethical justification for current or expanded use. Autonomy of parental decision making and expanded access favor the current and expanded use of NTD screening; whereas the probability of harm to aborted fetuses that would have had only mild handicaps had they been allowed to live argue for the contrary. We see no way of resolving this conflict. It must be delineated and the merits of both points of view debated. In the United States the result of such debate has thus far apparently favored the positive economic indications, improved autonomy of parental decision making, and improved access to services. As a result, expanded screening efforts have been funded.

VII. SUMMARY

We have attempted to delineate some of the economic and ethical consequences of screening for and intervening against several genetic diseases. If the proposed expansion of a screening program would result in positive economic and ethical consequences, it certainly should be recommended for expansion. When the consequences are in conflict, then the proper recommendation is not obvious and the trade-offs need to be delineated to facilitate discussion and debate.

For Tay-Sachs Disease screening, the case is quite straightforward, as both the economic and ethical consequences were positive. These programs clearly should be supported. With respect to Down's Syndrome, the economic consequences are positive; however, there is some question with respect to the ethical consequences. In our opinion the ethical consequences of screening programs for Down's Syndrome are positive. To the extent that social values do not favor the availability of abortion services for women pregnant with a Down's Syndrome fetus, our position would have to be qualified. However, we believe that social values do favor the availability of such services and that current screening efforts should be supported and access to services should be expanded for all pregnant women aged 35 and over. Furthermore, serious consideration should be given to extend these services to pregnant women aged 33 and 34.

Screening for Neural Tube Defects presents a more difficult case. Here the economic consequences of screening are positive, yet the ethical choice is not clear. We have offered no resolution to the ethical problem, but rather have attempted to delineate the nature of the trade-offs in an effort to facilitate discussion and debate. Actions such as the new statewide NTD screening program in California indicate that positive economic outcomes, improved autonomy of parental decision making, and improved access, at the expense of the lives of a small number of fetuses that would have been only mildly handicapped by the disease, is justified.

There are many genetic diseases for which screening tests and interventions have yet to be developed; they will have different economic and ethical consequences. We chose only three that we believe allowed a reasonable spectrum of discussion. However, in all three of these diseases, the economic indications for screening were positive. But what if they were not? Positive ethical consequences may still prevail. For the economic indications to be positive, there, at the very least, needs to be identification of an adequate number of cases and intervention in an adequate proportion of those. For several years the state of Texas required mandatory screening of all newborns for homocystinuria, a genetic disease which is caused by a deficiency of the enzyme serine dehydratase. This was done for the welfare of the affected infants. It thus had an ethical motivation. However, the incidence of this disease was so low that the economic criterion could not be met. Although there was a conflict between negative economic and positive ethical indications, the program was adopted. Interestingly, after several years, this screening program was discontinued (although mandatory screening of newborns for several other genetic diseases with higher incidence rates continues).

School of Public Health
University of Texas Health Science Center
Houston, Texas

and

School of Medicine
Harbor-U.C.L.A Medical Center
Los Angeles, California

NOTES

* The authors would like to thank Professor Daniel Wikler of the Program in Medical Ethics, University of Wisconsin Medical School, for his helpful comments.

[1] Unfortunately, about 15 percent of the U.S. population is uninsured and high out-of-pocket costs assuredly affect their decisions whether or not to seek care. What is true is that attention to the cost factor is a new concern for middle and upper income segments of the population. The authors are indebted to Daniel Wikler for this point.

[2] See ([3], [10], [33], [40]) for discussion of society's need to consider rationing. Cost-containment efforts that reduce inefficiencies in the provision of health care services do not reduce benefits to the patient; they just lower costs. Therefore, these policies are highly desirable, but they are not our primary concern as the attainment of ethically satisfactory outcomes need not be hampered by their implementation [4]. For a discussion of the evidence of inefficiencies in medical care, see ([32] pp. 185–190). With the continued aging of the population and advances in medical technology, it is unlikely that the difficult task of removing inefficiencies in the health care system alone will satisfy cost containment objectives. It is the possibility of reductions (or smaller increases) in the provision of *beneficial* health care services, in an effort to contain costs, that involves serious economic/ethical trade-offs.

[3] A useful exception is found in the President's Commission's discussion of the ethical and economic trade-offs that exist with respect to the exclusion of women under 35 years of age from amniocentesis:

> One concern is that sudden less restricted access to amniocentesis might have the effect of overwhelming the existing capacity for performing this procedure, with the result that some of the women who have the greatest need would fail to receive the test while those at lower risk do have it. Thus *it is important that the elasticity of the capacity for amniocentesis is studied* .... Moreover, amniocentesis is a costly procedure; it may not be efficient or equitable in light of other demands on scarce resources to expend public funds for groups at low risk . . . ([33], p. 81).

This is precisely what is needed — the willingness to openly discuss the trade-offs and in this particular case to identify issues that need further study. Only in this way can informed debate be made possible.

[4] Efficiency clarifies the relation between economic costs and benefits associated with a program, but it is not intended to be the sole decision-making criterion and should not be construed as such. However, while most health programs are undertaken for humanitarian reasons and thus to address ethical considerations, that does not mean that the efficiency aspects are without importance as resources are scarce and all wants cannot be fulfilled. As stated in *Genetic Screening: Programs, Principles, and Research*:

> ... the (CBA) perspective is essential to a systemematic evaluation of new opportunities for organizing, structuring, producing, and delivering all human services. The need for such activities reflects the basic economic assumption of scarcity of resources, an assumption well supported by the experience in the area of human services delivery. Given the scarcity of resources, it is essential that services be provided in ways that are most likely to generate positive benefits both for the users who avail themselves of those services and for society as a whole ([25], p. 212).

[5] The population should be identified by demographic characteristic, that is, maternal age or ethnic group, family history, prior pregnancy history, or identified through carrier screening such that the risk of occurrence is relatively high.

[6] It should be reiterated that we believe that the decision to abort a fetus affected with a serious genetic disease is one of conscience made by individual couples. Certainly others may disagree and give the abortion issue overriding weight and arrive at a different conclusion. This is certainly the subject of current social debate.

BIBLIOGRAPHY

[1] Adams, M., *et al.*: 1981, 'Down's Syndrome: Recent Trends in the United States', *Journal of the American Medical Association 246*, 758–760.

[2] Althouse, R. and Wald, N.: 1980, 'Survival and Handicap of Infants With Spina Bifida', *Archives of Disabled Children 55*, 845–850.

[3] Aaron, H. and Schwartz, W.: 1984, *The Painful Prescription: Rationing Hospital Care,* The Brookings Institution, Washington, D.C.

[4] Bailey, M.: 1984, 'Rationing and American Health Policy', *Journal of Health Politics, Policy and Law 9,* 489—501.

[5] Carter, C. and Evans, K.: 1973, 'Spina Bifida and Anencephalus in Greater London', *Journal of Medical Genetics 10,* 209—234.

[6] Center for Disease Control: 1978, 'Mental Retardation, Birth Defects and Genetic Disease Control Programs: A Cost-Benefit Analysis', Atlanta.

[7] Chamberlain, J.: 1978, 'Human Benefits and Costs of a National Screening Programme for Neural Tube Defects', *Lancet 1,* 1293—1296.

[8] Conley, R. and Milunsky, A.: 1975, 'The Economics of Prenatal Genetic Diagnosis', in A. Milunsky (ed.), *The Prevention of Genetic Disease and Mental Retardation,* W. B. Saunders Co., Philadelphia.

[9] Crandall, B., *et al.*: 1983, 'Maternal Serum Alpha-Fetoprotein Screening for the Detection of Neural Tube Defects', *Western Journal of Medicine 138,* 524—530.

[10] Fuchs, V.: 1984, 'The Rationing of Medical Care', *New England Journal of Medicine 311,* 1572—1573.

[11] Hagard, S., *et al.*: 1976, 'Screening for Spina Bifida Cystica: A Cost-Benefit Analysis', *British Journal of Preventive and Social Medicine 30,* 40—53.

[12] Hall, J. *et al.*: 1978, 'The Frequency and Financial Burden of Genetic Disease in a Pediatric Hospital', *American Journal of Medicine Genetics,* 417—436.

[13] Hook, E.: 1978, 'Spontaneous Deaths of Fetuses with Chromosomal Abnormalities Diagnosed Prenatally', *New England Journal of Medicine 299,* 1036—1038.

[14] Hook, E. and Chambers, G.: 1977, 'Estimated Rates of Down Syndrome in Live Births by One-Year Maternal Age Intervals for Mothers Aged 20—39 in a New York State Study: Implications of the Risk Figures for Genetic Counseling and Cost-Benefit Analysis of Prenatal Diagnosis Programs', *Birth Defects: Original Article Series 13,* 127.

[15] Hook, E. and Fabia, J.: 1978, 'Frequency of Down Syndrome in Livebirths by Single Year Maternal Age Interval: Results of a Massachusetts Study', *Teratology 17,* 223—228.

[16] Hsu, L.: 1978, Personal communication.

[17] Janerich, D.: 1972, 'Anencephaly and Maternal Age', *American Journal of Epidemiology 95,* 319—326.

[18] Layde, R., *et al.*: 1979, 'Maternal Serum Alpha Fetoprotein Screening: A Cost-Benefit Analysis', *American Journal of Public Health 69,* 566—573.

[19] Leck, I.: 1974, 'Causation of Neural Tube Defects: Clues From Epidemiology', *British Medical Bulletin 30,* 158—163.

[20] Lemire, R., *et al.*: 1975, *Normal and Abnormal Development of the Human Nervous System,* Harper, New York.

[21] Lubs, H.: 1977, 'Frequency of Genetic Disease', in H. Lubs and F. de la Cruz (eds.), *Genetic Counseling,* Raven Press, New York.

[22] March of Dimes: 1983, *Genetic Services, International Directory,* 7th edition, MOD Birth Defects Foundation, White Plains, New York.

[23] Massarik, F. and Kaback, M.: 1981, *Genetic Disease Control, A Social Psychological Approach,* Sage Publications, Beverly Hills, California.

[24] Milunsky, A.: 1973, *Prenatal Diagnosis of Hereditary Disorders,* Charles C. Thomas, Springfield, Illinois, 1973.

[25] National Academy of Sciences, Committee for the Study of Inborn Errors of Metabolism: 1975, *Genetic Screening: Programs, Principles, and Research,* National Academy of Sciences, Washington, D.C.
[26] National System of Hemophilia Treatment Center, personal communication.
[27] National Institute of Child Health and Human Development: 1979, *Antenatal Diagnosis: Report of a Consensus Development Conference,* NIH Pub. No. 80-1973, Bethesda, Maryland.
[28] Nelson, W., *et al.*: 1978, 'An Economic Evaluation of a Genetic Screening Program for Tay-Sachs Disease', *American Journal of Human Genetics 30,* 160—166.
[29] Nelson, W., *et al.*: 1978, 'A Comment on the Benefits and Costs of a Genetic Screening', *American Journal of Human Genetics 30,* 663—665.
[30] O'Brien, J.: 1970, 'Discussion of Massachusetts Metabolic Disorders Screening Program', in M. Harris (ed.), *Early Diagnosis of Human Genetic Defects: Scientific and Ethical Considerations, Fogarty International Center Proceedings, No. 6,* U.S. Government Printing Office, Washington, D.C.
[31] Powledge, T. and Fletcher, J.: 1979, 'Guidelines for the Ethical, Social, and Legal Issues in Prenatal Diagnosis', *New England Journal of Medicine 300,* 168—172.
[32] President's Commission for the Study of Ethical Problems in Medicine and Biomedical and Behavioral Research: 1983, *Securing Access to Health Care Volume One: Report,* U.S. Government Printing Office, Washington, D.C.
[33] President's Commission for the Study of Ethical Problems in Medicine and Biomedical and Behavioral Research: 1983, *Screening and Counseling for Genetic Conditions,* U.S. Government Printing Office, Washington, D.C.
[34] Prest, A. and Turvey, R.: 1965, 'Cost-Benefit Analysis: A Survey', *The Economic Journal 75,* 683—735.
[35] Sadovnick, A. and Baird, P.: 1983, 'A Cost-Benefit Analysis of a Population Screening Programme for Neural Tube Defects', *Prenatal Diagnosis 3,* 117—126.
[36] Special Report of a Consensus Meeting: 1985, 'Maternal Serum Alpha-Fetoprotein Screening for Neural Tube Defects', *Prenatal Diagnosis 5,* 77—83.
[37] Swanson, T. E.: 1970, 'Economics of Mongolism', *Annals of the New York Academy of Science 171,* 679—682.
[38] Swint, J. M., *et al.*: 1979, 'The Economic Returns to Community and Hospital Screening Programs for Genetic Disease', *Preventive Medicine 8,* 463—470.
[39] Tetrault, S. and Scott, R.: 1974, 'Urban Hospitalization Cost Analysis of Patients With Sickle Cell Disease', Unpublished paper, Howard University; abstracted in the *Proceedings of the First National Symposium on Sickle Cell Disease,* DHEW, NIH Publication No. 75-723, Washington, D.C.
[40] Thurow, L.: 1984, 'Learning to Say No', *New England Journal of Medicine 311,* 1569—1572.
[41] U.S. Department of Health, Education and Welfare, Public Health Service, National Institute of Health: 1979, *Antenatal Diagnosis,* U.S. Government Printing Office, Washington, D.C.

STUART F. SPICKER

ETHICAL REFLECTIONS IN GENETIC SCREENING: A REPLY TO SWINT AND KABACK

I. INTRODUCTION

Before replying to the contribution of J. Michael Swint and Michael M. Kaback, in their "Intervention Against Genetic Disease: Economic and Ethical Considerations," it may be useful to consider their examination of the economic and ethical consequences of three genetic screening programs in the wider context of a prominent *dilemma* which is presently confronting the public, virtually all health professionals, the United States government, and even the entire market economy ([2], [3]). In the present context, where we have been asking what, if anything, is the price of health, it may be useful to formulate the central predicament which is now upon us, and to appreciate the fact that two equally distasteful conclusions or outcomes are about to be realized.

Premise 1: *If* the U.S. persists in its past economic policy with regard to health care expenditures and thus continues to increase the funds expended on the health care requirements of its citizens (we already expend almost 11 percent of the G.N.P. on health care), *then* national bankruptcy or other severe economic consequences will ensue; *and, if* we do not continue to provide the economic resources and material support required for health care (and by so doing maintain a high quality of health care), *then* we shall suffer a serious decline in the health status of our population as a whole, and particular patients will undoubtedly suffer great burdens such as pain, chronic illness, and untimely death.

Premise 2: *Either* we must establish a health policy which infuses additional funds and resources toward the health care of the citizenry, *or* we must dramatically restrict, reduce, and even ration the resources needed to maintain a high quality of health care.

Q.E.D.: We shall *either* suffer severe economic consequences like national bankruptcy and/or an increased federal deficit, *or* the citizenry will suffer a decline in its health status, and particular patients will be compelled to endure unnecessary pain, chronic illness, and even untimely death.

G. J. Agich and C. E. Begley (eds.), The Price of Health, 157–164.

Formulating the critical problem this way can reveal and highlight the tension that exists between ethics and economics, or (to be more precise) between ethical principles and economic propositions. As is well known, the two chief ways to avoid the unpleasant alternatives of the conclusion noted above are to show (1) that the second premise is false — that the disjunctive ("either/or" where "or" is taken exclusively) is false. If one can show this, then one has shown that the particular dilemma is not adequate grounds for accepting either of the two unpleasant conclusions, though the conclusion may still be true; or, one can show (2) that the first premise is to be rejected by showing that health spending poses no threat to the nation's economic well-being.

Swint and Kaback have not expressed their position in precisely this form, nor have I set it out to make my own case for the inevitability of the bankruptcy of the U.S. Treasury or, equally unpleasantly, to make a case for a rapid decline in the nation's health status and thus a concomitant and rapid increase in human suffering. I simply wish to illuminate the viewpoint which is presupposed by the authors, as well as many others: that genetic screening programs (in this case) which demand our nation's resources and funding bring about greater benefits than burdens. For this reason, Swint and Kaback have no serious concern about the costs of such programs on the one hand and the quality of the nation's health and economic status on the other. From their vantagepoint, we should pay for those screening programs which are cost-beneficial unless they (or any one of them) fail to be morally justified. The question which they do not address, however, is: Should genetic screening programs be supported when they meet appropriate economic criteria *and* pass the test, so to speak, for our moral approbation, or at least do not cause harm? [1]

II. THE CRITICAL PREMISES

Of the three general types of screening programs now extant: postnatal screening of newborns, ante- or prenatal screening on pregnant women, and screening for so-called "carriers" of deleterious genes, Swint and Kaback confine themselves to the second and third: Down's Syndrome and Neural Tube Defects by prenatal detection, and Tay-Sachs Disease by mass screening of adults of child bearing age, especially persons of Ashkenazic Jewish descent.

Swint and Kaback embrace a series of premises and assumptions in developing their position which are worth noting:

1. There exists today a scarcity of and burden on economic resources, both human and material, given the competing demands upon limited health care dollars. The predominantly genetic diseases of human fetuses represent only one area where dollars are needed; there are, of course, others like the future prospects of natural and artificial heart transplantation for patients of all ages – neonates as well as the elderly [4]. Hence, with regard to the provision of health services and the increased demand for medical resources it has become necessary to: (a) reduce the overall medical/health costs (often crudely reflected in the present and rising GNP, now at almost 11 percent); (b) avoid "production losses" such as the additional cost when workers are removed from the labor force due to illness, or the enormous time and material required for children and others who are chronically ill; (c) drastically reduce expensive institutionalization; and (d) reduce, wherever humanely possible, the costs of special education and training.

2. Swint and Kaback argue that it is essential to assume *scarcity* of resources, an assumption, even if challenged with regard to the present, will surely be true in the not-too-distant future. Hence, it is a prudent assumption for viewing the conflict between ethical principles and economic propositions regarding health care.

3. The authors are appropriately wary of the advisability of letting techniques such as cost-benefit analysis (CBA) and cost-effectiveness analysis (CEA) determine medical and clinical decisions that bear on life sustaining interventions. Hence, CBA and CEA methodologies are regarded as merely a "decision-assisting technique" which necessarily "disregards" ethical analysis.

4. The authors tend to equate "ethical considerations" with that which is beneficial or with that which prevents harmful outcomes. "Unethical" tends to be identified with *harmful acts and choices.* Swint and Kaback argue that we should try to reduce pain and suffering wherever possible. Furthermore, they think that economic considerations taken in isolation cannot in themselves justify ethical propositions. They strive to have all economic decision making in the health care arena cohere with the principle of utility: the view that the rightness or wrongness of a policy depends only on the total goodness or badness of its consequences or states of affairs brought about by the behavior or

action, that is, the effect of the behavior on the welfare of all human beings.

III. SCARCITY, RATIONING, AND THE "GOODS" OF HEALTH CARE

Swint and Kaback call our attention to the issues which must be considered in developing defensible standards of rationing. As is patently obvious, rationing of health care has always occurred, as have decisions to terminate treatment. Furthermore, rationing is likely to continue since, as our authors affirm, no country in the world can provide state of the art medical care for all its citizens. Yet as one senior vice president and treasurer of a Connecticut Hospital remarks: "We're not sure what the long-term impact of the new regulations will be. We may have to manage our resources differently. But we will always meet our responsibilities to the community. If people are sick, they will get the level of care they need. That is why Mount Sinai Hospital [Hartford, Connecticut] exists" ([8], p. 9). Do such pronouncements really convince? Do they jibe with the present economic reality and trends? Swint and Kaback think not; and I agree.

The authors support the methods of CBA and Cost/Cost Analysis in order to determine the demands on the marketplace and to be better informed in order to make the most rational determinations regarding that which bears directly on society's health. Objections regarding the use of CBA and CEA to determine the value of human life have already been raised in this volume. Suffice it to say that there are major difficulties with interpreting from a CBA or a CEA what we *ought* to do with respect to various health care problems, especially in cases in which distributive justice is desired while rationing is a pertinacious reality. Nonetheless, Swint and Kaback apparently seek to reconcile or balance principles of equity (or ethics) and efficiency in health care, especially as they focus on genetic disease. The conflict might be formulated as follows: there is a need to affirm the fundamental social value of saving lives while balancing that against other claims on scarce resources.[2] But here we encounter a common fallacy in the literature that argues that we must "balance" A against B, make a "trade-off" between A and B.

The notion of achieving a "balance" in the context of equality and efficiency, or ethics and economics, is misleading. The metaphor of balancing suggests an objective scale — a pair of trays, each tray revealing an objective determination thanks to gravitational force. By utilizing the "balancing" or trade-off metaphor, what is highly subjective and even normative is made to masquerade under the guise of something objectively determinable and descriptive. But there is nothing that functions as neutrally as gravitational force in the economics/ethics problematic. Hence, the notion of balancing is unhelpful here and raises more questions than it provides solutions.[3]

Swint and Kaback proffer the following thesis: Scarcity of resources generates difficult economic and ethical trade-offs such as new forms of rationing [6]. They argue: ". . . our wealth has in many instances [until quite recently] allowed us the nearly exclusive use of ethical and medical criteria for making explicit health care utilization decisions. But today we are faced with economic conditions which cause us to radically revise the basis on which these utilization decisions are to be made, as well as the type of decisions themselves."

They illustrate the pressure that economic scarcity has produced by cost benefit analysis of screening procedures for Tay-Sachs Disease, Down's Syndrome, and Neural Tube Defects. The questions posed by a study of the costs and benefits of screening for these diseases are, first, whether the screening programs are economically worthwhile, that is, are the benefits to be gained from knowing quite early which infants are at high risk for the disease in question outweighed by the costs of the screening programs themselves? Second, are the screening programs morally or ethically justifiable, that is, do they compromise or harm those who participate in them? In the cases of Tay-Sachs Disease, Down's Syndrome, and Neural Tube Defects, are pregnant women or embryos (fetuses) compromised or harmed in ethically unacceptable ways? I do not find it useful here to argue whether or not these screening programs — on-going or proposed — sustain the various moral concerns mentioned by the authors, since this would require an extensive sojourn into moral philosophy and bioethics. I will let stand the claims made by the authors, namely that these ethical concerns are in general not compromised by the genetic disease screening programs which they have described. I shall attend, instead, to the first issue: the cost-benefit determination of the screening programs themselves.

Swint and Kaback argue that all three screening programs are indeed cost effective. There is, generally speaking, a positive cost-benefit ratio that economically warrants conducting these screening programs. But could their conclusion have been otherwise? It hardly seems likely that it could.

The empirical information required to determine the screening *costs* is, though not easy to compile, certainly calculable, and experts in CBA are typically competent in determining this total cost picture. But the true cost-benefit assessment is far more complex. In this context, Swint and Kaback should have compared the cost of raising a normal, non-affected infant against the additional costs of bearing and raising infants who manifest any one of these genetic diseases. Given that Tay-Sachs infants do not live beyond four to five years, this assessment of cost is calculable, even if it does not include the costs of the psychological trauma of raising a child to age four or five. The Down's Syndrome (Trisomy 21) infants do indeed live much longer than the Tay-Sachs infants, and the additional costs for raising these children (given a range of disability from mild to extreme retardation) is surely relevant to society in comparison to the costs of raising non-affected newborns. Analogous points apply to the Neural Tube Disorders which lead to extremely expensive medical care costs.

A peculiar fact clearly emerges when we consider these three genetic diseases: the "benefit" factor is essentially (but not totally) determined by comparing the costs of raising a child for an average expected lifetime versus the costs of terminating the pregnancy by abortion (as even the President's Commission has observed) ([9], p. 85). Swint and Kaback, then, have really compared screening to abortion intervention. They begin to signal the problem, though they do not confront the so-called "abortion debate" in detail; however, they do tacitly ask: When is abortion ethically justified even from the perspective of the unborn?

Swint and Kaback's conclusions with regard to the economic analyses of these three genetic disease screening programs, then, rest squarely with comparing costs of bearing and raising these newborns and all later child care *against* the costs of voluntary abortion. To avoid the abortion issue, however, is to disregard one of the central ethical conflicts generated by these screening programs.

I am reminded of Woody Allen's one-liner: "Death is a great way to cut down expenses". It is a humorous cliché, of course, precisely because it is true. Hence, if abortion *with non-replacement*[4] of the affected fetus by an unaffected one at a later date is the chief cost-effective factor in determining the costs-benefit ratio of screening programs for the genetic diseases described by Swint and Kaback, then what has been said beyond the propositions that (1) "normal" life is more economical overall than "abnormal" life, and (2) non-life is more economical than enduring life itself — assertions equally true as they are obvious. The more important calculation requires the researcher to determine the cost of raising essentially healthy infants *against* the cost of caring for children with deleterious genes who live on because abortion was *not* practiced.

If no one intended to abort affected fetuses, of course, screening programs would have to be justified for purposes other than those typically mentioned today. If one could leave the abortion issues aside, then one could surely concur with Swint and Kaback that screening programs for the three genetic diseases discussed by them are cost effective, that is, economically efficient programs. The problem, however, is that even if *efficiency* and *social benefit* are determined by external evaluations of the lives of neonates with deleterious genetic factors in comparison with the costs of bearing and rearing unaffected fetuses; efficiency, like all so-called trade-offs, is discovered to be a *normative* concept masquerading as a straightforward and quantifiable *descriptive* state of affairs.

University of Connecticut Health Center
School of Medicine
Farmington, Connecticut

NOTES

[1] See [7], [10]. It is important to note that Swint and Kaback somewhat indirectly claim that the three genetic screening programs which they assess are weighed against the principle of beneficence. This claim requires that such genetic screening programs (a) prevent harm, (b) remove an evil, and even (c) promote good ([1], p. 107).
[2] See the discussion in the Report of the President's Commission where "balancing of benefits and harms" is mentioned in the interests of equity ([9], p. 84).
[3] There can be little dispute with the authors that, due to a significant rise in average life expectancy, persons with late onset genetic disease will incur increasing health care needs. Indeed, it should be stressed that many people tend to confuse "life span" with "average life expectancy"; it is the latter that has been increasing in this century. The human "life span", on the other hand, is actually little changed from the time of our ancient ancestors. The important point is that more and more people are living fourscore and ten, thus increasing dramatically the average life expectancy and thus the prevalence of late-onset genetic diseases. Nonetheless, such projections are virtually

useless in determining what we *ought to do.* We can, for instance, properly criticize the United Kingdom for failing to initiate their subtle and tacit policy of "incrementalism" with regard to hemodialysis for ESRD on the basis of a comprehensively conducted CBA. Thomas Halper quite correctly has pointed out that the actual "selectors" for treatment or non-treatment for ESRD are the general practitioners, since they serve as gatekeepers for renal hemodialysis treatment [5]. Unfortunately, their decisions are binding and are not made outside the clinic, as perhaps they should be.

[4] The "replacement case": the voluntary aborting of an affected fetus followed at later life by the birth of a non-affected fetus from the same woman. The "non-replacement case": the voluntary aborting of an affected fetus followed by no future birth by the same woman.

BIBLIOGRAPHY

[1] Beauchamp, T. L. and Childress, J. F.: 1983, *Principles of Biomedical Ethics,* 2nd edition, Oxford University Press, New York.

[2] Enthoven, A. C.: 1978, 'Consumer-Choice Health Plan: Inflation and Inequity in Health Care Today', *New England Journal of Medicine 298,* 650—658.

[3] Enthoven, A. C.: 1978, 'Consumer-Choice Health Plan: A National-Health Insurance Proposal Based on Regulated Competition in the Private Sector', *New England Journal of Medicine 298,* 709—720.

[4] Graven, D., *et al.*: 1984, *The Price of Life: Ethics and Economics: Report of the Task Force on the Affordability of New Technology and Highly Specialized Care: Life at Any Price?,* Minnesota Coalition on Health Care Costs, Minneapolis, Minnesota.

[5] Halper, T.: 1985, 'Life and Death in a Welfare State: End-stage Renal Disease in the United Kingdom', *Milbank Memorial Fund Quarterly 63,* 52—93.

[6] Heckler, M. M., *et al.*: 1985, 'Ethics, Rationing and Economic Reality', *Federation of American Hospitals Review 18,* 14—43.

[7] Hilton, B., *et al.*: 1973: *Ethical Issues in Human Genetics,* Plenum Press, New York.

[8] *Mount Sinai Hospital Annual Report:* 1984, 'No Easy Answers: The Issue of Medical Ethics', Mount Sinai Hospital, Hartford, Connecticut.

[9] President's Commission for the Study of Ethical Problems in Medicine and Biomedical and Behavioral Research: 1983, *Screening and Counseling for Genetic Conditions: A Report on the Ethical, Social, and Legal Implications of Genetic Screening, Counseling, and Education Programs:* U.S. Government Printing Office, Washington, D.C.

[10] Twiss, S. B.: 1974, 'Ethical Issues in Genetic Screening: Models of Genetic Responsibility', in Bergsma, *et al.* (eds.), *Ethical, Social and Legal Dimensions of Screening for Human Genetic Disease,* Stratton Intercontinental Medical Book Corp., New York, pp. 220—261.

MARY ANN BAILY

RATIONING MEDICAL CARE: PROCESSES FOR DEFINING ADEQUACY

I. INTRODUCTION

For most people, health care is special because of its importance in preventing pain and suffering, preserving the ability to pursue a normal life plan, providing information and relieving worry, and reflecting a community's concern for its members. As a result, ensuring access to health care is an important goal for modern societies. Translating this goal into practical policy, however, has proved more difficult than accepting it in principle.

In the health policy literature, three implicit concepts of equitable access are current ([21], [35]). One is based completely on patient need ([3], [10], [23], [58], [59]). A second emphasizes market outcome [6]. The third, a compromise between the other two, gives special priority to a basic level of care (a "decent minimum" or "adequate level"), leaving care above this level to be handled by markets.[1] In the present paper I am concerned primarily with this latter concept of a decent minimum or adequate level of health care.

The decent minimum approach commands considerable support, whether on pragmatic grounds or out of philosophical conviction.[2] Nevertheless, there is difficulty in making it operational. The difficulties are of three kinds. First, there is no consensus on the specification of the decent minimum. The care actually available to those who cannot afford to pay varies arbitrarily along dimensions unrelated to health condition. It is unclear how to develop a principled consensus on the content of the decent minimum in a pluralistic society such as the United States.

Second, this approach conflicts with physicians' current perceptions of their ethical duty. A delivery system based on a decent minimum requires limits to be set on care in accord with cost. In the United States, many physicians feel it is unethical to consider cost in their clinical decisions, let alone to set limits on the care provided that vary with income and source of payment ([2], [4], [41], [42], [50]).

Third, there is no consensus on how much patients must be told

G. J. Agich and C. E. Begley (eds.), The Price of Health, 165–184.

about the limits on care imposed to conserve resources. The paternalistic model of the physician-patient relationship has given way in recent years to a model of shared decision making in which patients receive much more information about treatment options than in the past. But this makes it more difficult for the provider to act as the gatekeeper, refusing resources the patient wants.

In this paper I discuss these barriers to implementation of a decent minimum. I argue that they are difficult but not impossible to surmount and suggest some strategies for doing so.

II. DEFINING THE DECENT MINIMUM

In seeking a concrete specification of the decent minimum, it is natural to look for guidance in the philosophical arguments advanced in its favor. However, I do not offer a critical evaluation of the competing theories. Instead, I focus on the process of translating the decent minimum approach into practical policy rather than justifying it. In a pluralistic society, no one ethical theory is likely to become dominant. Although the underlying theories are very different, they do in fact have practical implications in common, and so their differences are less important in practice than they at first appear. Thus, considerable policy insight can be derived from the philosophical arguments even without choosing among them. For example, all the arguments imply that the importance of health care — the extent to which it is special — depends on an individual's health condition. Therefore, a decent minimum cannot be simply an amount of money or a fixed amount of care; it must be defined in relation to health state. Any system to deliver the basic minimum must be able to vary the care according to individual need.

All philosophical arguments imply that the definition of adequacy should depend on the availability of resources. The benefits of health care must be weighed against costs in the light of competing uses for resources. Specification of adequate care for a particular health condition requires specification not only of the kind and amount of care but also of the quality, which will not necessarily be the highest possible. For some conditions, adequacy may even mean no treatment at all.

This point means that defining adequacy requires detailed information on medical technology and on preferences — how health benefits can be produced and how society regards the trade-offs among different kinds of benefits obtainable from health care and between health benefits and other kinds of benefits. Since technology, preferences, and resource availability are changing constantly, the specification of an adequate level must be capable of constant revision. Thus, it is incorrect to assume, as many people do, that the decent minimum approach implies a once and for all decision which is then imposed on the delivery system. Rather, what is needed is an ongoing *process* capable of incorporating changing information on technology, preferences, and resource availability.

Because adequacy is dependent on health condition, the process requires the active cooperation of providers. They will have to evaluate the health state of their patients and their clinical decisions will have to reflect the current specification of the adequate level standard. Since that specification takes cost into account as well as benefits, individual provider ethics must allow doctors to limit beneficial care. In other words, they must be allowed to ration.

The process must be linked to the political process. In the United States, this is an obvious way to incorporate societal preferences. More importantly, guaranteeing access to an adequate level of care for the poor requires the coercive power of government to ensure availability of funds.

There are also differences among the philosophical arguments which should be noted. A major difference is in the importance ascribed to the claim of personal health care on society's resources. For some, individuals have a right to obtain a decent minimum; for others, there is no right but society has an obligation to provide it. For the rest, there is neither a right nor an obligation, but to provide a minimum of health care would be good. Indeed, it would be a moral outrage if society did not provide at least a basic minimum of care to the needy.

This difference is of less practical significance than it seems at first. All three positions imply reliance on government to pursue a collective goal. The rights- and obligation-based arguments provide stronger justifications for coercive government action, such as interference in individual consumption plans. However, in a society that readily accepts taxation to provide public libraries, recreational facilities, highways, and the like — the absence of which would *not* be a moral outrage — the last position seems strong enough. Note that the strength of the claim does not bear on its extent.

A second difference among arguments lies in the differing priorities implied in specifying exactly what is guaranteed. The equal opportunity formulation emphasizes care that enables a restoration to normal functioning. A utilitarian formulation might put more emphasis on the relief of pain and suffering even if there were no possibility of return to normal functioning. A contractarian formulation based on Rawls' maximization of the least advantaged, the so-called *maximin* principle, might give more care for rare conditions than a utilitarian formulation.

Again, the difference among theories is of less practical significance than it appears. For practical policy, the difficulty is not that the adequate level is characterized differently by the different approaches, but that these approaches do not provide a concrete characterization of adequacy at all ([4], [15]).

III. ADEQUACY AS A STANDARD OF CARE ACCEPTABLE TO THE MIDDLE CLASS

How then to set the level? I suggest that in the United States adequacy should be specified as a standard of care that is acceptable to the middle class. In other words, the floor should be such that when that level of care is guaranteed, a person of average income feels no necessity to purchase additional care, even though more is available at its social cost. This would mean that the majority of the population, including the poor, would receive a common standard of care.[3] Equity of access would be defined in terms of equality relative to the level of services consumed by the middle class.

Such a common standard, however, is not required by the logic of the decent minimum. The decent minimum concept of equity of access does not preclude, but rather supports two-class — even multi-class — systems, since incomes and preferences differ and people who want more than adequate care are permitted to buy it at social cost. However, setting a middle-class standard, under the right conditions, would overcome some practical difficulties in implementing the decent minimum and serve as a reasonable approximation to the level of care morally required.

Of course, it might be objected that such a system would be prohibitively expensive. After all, the appeal of the decent minimum approach for many is the permission it gives to abandon the goal of providing expensive "mainstream" medicine to the poor. It is thus important to emphasize that I am not proposing that the standard of care *currently* received by the middle class be extended to all. There is nearly universal agreement that this standard is wasteful and inefficient. In the existing delivery system, there is no effective mechanism for ensuring that appropriate account of cost is taken in the making of clinical decisions. Both consumers and providers face perverse incentives; as a result, the standard of care now delivered to the average middle class person does not represent the standard such a person would choose if the costs were reflected in actual health insurance premiums and taxes ([7], [25], [26]).

A decent minimum approach requires doctors to limit beneficial care. Setting one basic standard of care as normal practice (with the understanding that there would be standards above that) would be far easier for the medical profession to accept than setting a normal standard for the average patient and an explicitly lower standard for the poor.

Doctors resist the role of rationer of societal resources. Of course, the role is often thrust upon them, nonetheless. The apportionment of scarce resources among patients with competing needs is most apparent when there is a physical shortage of resources such as intensive care beds, dialysis machines, or donated organs, but in fact, limits on beneficial care are common. For example, most doctors would acknowledge that they allocate their scarce time among patients and that some patients would benefit from more physician time spent on histories and careful examination [47]. However, doctors prefer not to see their activities as rationing.

A fair rationing system is actually in the patient's interest since patients ultimately pay the cost of low benefit care in higher insurance costs. If the doctor's ethical obligation includes serving as agent in the economist's sense of acting in the patient's best interests given the patient's preference and financial situation, then it is ethical for the doctor to participate actively in rationing. Of course, if the patient is not paying out-of-pocket at the time clinical decisions are made, the patient will have an incentive to subvert the system, but at least there should be no ethical conflict for the physician in upholding it, under this interpretation.

Many people prefer the ethic of the doctor as patient advocate — the ethic that the doctor's duty is to provide all beneficial care without regard to cost. Doctors find it difficult enough to balance patient good against their own good; acting as steward of society's resources seems far too difficult a task. Likewise, patients prefer to have their doctors acting in their best interest, rather than sacrificing their welfare to some concept of the social good ([41], [42]).

Unfortunately, there is no way to preserve clinical autonomy for doctors to respond to individual situations, allow them to practice an ethic dictating that they do everything possible for their patients (without regard to cost), *and* limit the care patients receive to that which is worth its cost. It has been suggested that the task could be accomplished by some combination of limits on resources and explicit rationing rules; within these restrictions, doctors then could be patient advocates ([27], [47]). But the complexity of medical technology and the variability in individual cases are so great that doctors can always sabotage the system if they choose unless there is very extensive and intrusive review of clinical decisions. A rationing system cannot function effectively unless doctors accept its legitimacy.

It is *possible* to have a system in which there are two rationing standards, one for the poor and another for the better off. Some developing countries, for example, have a public system of care in which a radically different standard of care is delivered than is delivered to private patients. In these countries, some doctors work in public clinics and in private practice, moving back and forth easily between the two standards. Most American doctors, however, have ethical difficulties with a standard of treatment that varies in medically important ways with a person's income or the source of payment. Therefore, it might be easier to obtain the cooperation of doctors if there were one standard of care for the majority of the population.

Would such a standard require a level of care that exceeds the public's sense of what is morally required? In other words, would the standard be unacceptably high? The evidence is mixed. On the one hand, the rhetoric surrounding public programs suggests strong suspicion of "two-class" medical care and institutions that serve only the poor. Medicare and Medicaid were designed to integrate the elderly and the poor into the mainstream of medical care. Aspects of care which are seen as amenities or relatively unimportant to outcome such as waiting time, amount of time spent with a doctor, or continuity of provider can vary in ways unacceptable to the average paying patient, but people do not seem to be comfortable with the idea that the actual standard of medical treatment should vary with a person's income or the source of payment. Medical malpractice law recognizes that standards of care may vary from one locality to another and between general practitioner and specialist; it does not recognize, however, a variation in standard of care based on the income or insurance status of the patient.

Of course, there is a gap between rhetoric and reality. Many people do not have access to care, and the actual standard of care provided to the poor does not match the rhetoric. Although public hospitals are supposed to deliver the same standard of care as other hospitals, many are not supplied with the funds to do so. Medicaid patients cannot find mainstream doctors to treat them because Medicaid reimbursement rates are so low.

In responding to this situation, the American public faces a dilemma. On the one hand, people believe that in the richest country in the world, every citizen ought to have access to good medical care. On the other hand, they have the uneasy sense that the standard of care now received by the middle class is wasteful and expensive, and not entirely satisfactory at that; yet, they believe that extending that standard of care to the poor will be prohibitively expensive and not worth the money in the benefits it provides.

It is instructive that people prefer to deal with this dilemma by persisting in believing in the face of evidence to the contrary that, with occasional unfortunate exceptions, everyone in this country already has access to good care. Once in a while the system fails, but normally people who really need health care receive it, and the health care they receive compares favorably with the care available to the average person. The more concerned people are about the size of government expenditures, the more unwilling they are to believe that serious deficiencies in access to care exist.[4]

If people did not have a moral intuition that adequate care should be available to all, and that adequacy means a standard of care acceptable to the average paying patient, this tendency would not be so strong. Thus, I think there is support for a standard of adequacy defined as that acceptable to the average person *if* the practical problem of the high cost of such a standard can be solved.

Those who do not believe the societal obligation is so extensive might be convinced to support the standard on other grounds [15]. Some groups may deserve health care in compensation for past injuries that have resulted in lower health status — minority groups or combat veterans. Other care may be justified on prudential grounds – social benefits from the greater productivity of a healthy work force or savings in other public programs. Finally, setting adequacy at a level that is satisfactory to the majority of the population lessens the danger that the floor will be allowed to fall too low because it is a separate level set for a politically powerless group. The history of institutions providing public services to low income or minority groups suggests that they are less responsive than institutions serving members of the broad middle group ([8], [33], [38], [39]).

It might be argued that the standard of care would not be too high but too low. This objection seems implausible for conditions that are reasonably likely to happen to the average person. By the very logic of the decent minimum, well-informed people of average means can be expected to want at least that level of care; it is difficult to see why they should be forced to guarantee themselves access to more than they want.

People may not be well-informed, however. Even if well-informed, they may be subject to systematic biases in their decision making. Setting the standard of care requires people to weigh probabilities — to decide how much it is worth spending on hypertension treatment to lower the risk of cardiovascular disease, to decide whether they want to pay to guarantee the availability of a heart transplant when the chances of needing one are very small, and so on. Decision making under uncertainty is difficult. Studies have shown that people have particular difficulty making decisions when there is a small probability of a very adverse event. Of course, many decisions in health care are of this type.

The fact that the probability of incurring many illnesses is not evenly distributed raises an additional problem. Those conditions that are unlikely to be experienced by the average person may receive less consideration than more common conditions of similar severity. Genetic and congenital conditions raise this problem in particularly clear form.

It is important to emphasize that the standard of care acceptable to the person of average income would be a reasonable first approximation to an adequate level, but it would not necessarily be fully satisfactory. People might well differ in their assessments of the extent to which it should be modified and in which direction. Unfortunately, the conditions that this approach would not handle well – for example, preventive care, genetic and congenital conditions, and organ transplants — are exactly the conditions about which the underlying philosophical theories differ. Even if a consensus about the amount of care required could be reached, I do not think it will be easy to assure that the system actually provides everything necessary, if it is not likely to be used by the average person. A variety of institutional mechanisms would be needed to deal with the questions and conflicts that would arise. These mechanisms might include special congressional committees, private charitable organizations, hospital ethics committees, and so on.

This debate, however, would take place within a system in which the broad outlines of adequacy were set. This would be an improvement over the existing situation. There would be an economizing of social attention. Questions such as whether prenatal care would be provided and in what amounts, how much cancer screening should be provided to the general population, and whether well-child care should be provided, would be settled and society could concentrate on questions such as how much care should be available for hemophiliacs.

IV. TWO EXAMPLES OF PROCESSES FOR DEFINING AND DELIVERING AN ADEQUATE LEVEL

Admittedly, there is still considerable indeterminacy in the approach recommended. What does it mean in practical terms to say "the floor should be such that when that level of care is guaranteed, a person of average income feels no necessity to purchase additional care, even though more is available at its social cost?" What is average income? How can the tradeoffs between health benefits and costs be observed without losing the benefits of risk-spreading through insurance? How can providers be brought to respect these tradeoffs in their clinical decisions? What percentage of care bought outside the system for a given condition triggers a re-evaluation of the level within the system? How is the cost of the system to be distributed so that no one's burden is excessive?

These questions could be answered in a variety of ways. My purpose here, however, is not to answer them, but to argue that they can be answered. In other words, I am only arguing that a practical system can be developed along the lines described. To demonstrate this thesis, consider two systems which have attracted special attention recently for their resource allocation process and the ethical acceptablility of the resulting distribution of care: a national health service such as that of Great Britain and a system of competing health plans such as Alain Enthoven's Consumer Choice Health Plan. The two systems are very different ideologically — socialized medicine versus market competition. Yet they are similar in that each constitutes a delivery system in which is embedded a process for defining and guaranteeing universal access to a standard of care which approximates adequacy.

The British National Health Service (NHS) has been examined in fascinating detail by Henry Aaron and William Schwartz [1]. They describe how the standard of care provided by the NHS is determined. What is interesting for this paper is the complexity of the process. It is often assumed that the decent minimum must be determined by a "central czar", or at least a "central committee", at a level far removed from the patient, and then imposed on the delivery system through inflexible regulation. This is not necessarily the case. Even in the "socialized" National Health Service, there is no single authority making all the decisions. Instead, a system of interlocking levels operates with different mechanisms at different levels.

At the top is a global budget for the entire health service determined by the Treasury and approved by the Cabinet and House of Commons — in other words, through a political process in which health care competes with other goods. At the bottom are the general practitioners, paid on a capitated basis and serving as gatekeepers to the system. In between are regional and district planning authorities and the hospitals. The regions receive fixed budgets which are then allocated by planning authorities including physicians and non-physicians. The hospitals also receive fixed budgets, which are allocated by the medical staff, composed of salaried physicians working directly for the hospital.

This complex process has the characteristics described as essential to defining and delivering adequate care. First, it is a process, not a static allocation, in which the level of care provided evolves with changing technology and societal preferences. Second, there is a political mechanism for determining the overall availability of resources to the sector and weighing health benefits against other uses of resources. Third, both physicians and non-physicians participate in the allocation process at the middle levels, where tradeoffs are made among various health benefits. And fourth, the use of physicians as gatekeepers at the individual patient level ensures that the care provided varies with health condition.

Viewed from this perspective, the NHS is certainly not perfect. Attempts to eliminate regional inequalities in the distribution of health care personnel and facilities have not been completely successful. Criteria for obtaining treatment vary from one area to another; for example, the criteria for selection for hemodialysis vary from center to center. Rationing of care falls disproportionately on care that requires expensive equipment, since it is easier to limit supply of such equipment and thus limit doctors in this way than it is to ration other kinds of care.

Nevertheless, the system seems to work. The small size of the private sector in medicine over the first three decades of the NHS's existence attested to the acceptability to the average Briton of the standard of care provided, although the private sector's spectacular recent growth suggests that the standard may no longer be as acceptable.

A different approach to health care delivery is the Consumer Choice Health Plan (CCHP) ([25], [26]). This approach leaves more choice to the individual consumer and relies less on bureaucratic decision making and control. Consumers are given incentives through the income tax to join health plans, so-called health maintenance organizations (HMOs). Each health plan is a kind of mini-health service, which sets a standard of care for its members. Competition among health plans leads to a distribution of standards of care that reflects consumer preferences. A system of government regulation is designed to ensure that the competition takes place along socially desirable lines — for example, that consumers are well-informed about the differences among health plans.

What makes this a method of setting an adequate level is the provisions for the poor and for others who would have difficulty obtaining health insurance coverage in a purely private market. These people are allocated purchasing power to enable them to join health plans. A complex set of regulations ensures that these people will belong to health plans that also serve middle class Americans and that they will receive a certain basic list of services. CCHP has been criticized for providing only a list of required basic services rather than specifying amounts in relation to health condition as well ([4], [15], [19]). A list of services clearly does not constitute a specification of adequacy — but in CCHP, the HMO includes a rationing agent and a set of incentives and regulations conditioning the behavior of the HMO.

As with National Health Service, Consumer Choice Health Plan embodies a process of defining adequacy in which the standard of care guaranteed to all can evolve over time. Physicians serve as gatekeepers, adjusting the care provided to individual patient need but operating within organizations that have financial incentives to weigh the benefits of care against the costs. Instead of relying on the political process, the preferences of consumers, backed by their freedom to change plans if dissatisfied, are expected to determine tradeoffs among health benefits and between health and other goods. The government's role is limited to maintaining the conditions necessary to allow consumers' preferences to influence the outcome and assuring that everyone has the resources to join a plan.

To be sure, this proposal has its weaknesses ([4], [21], [56]). Consumers may have difficulty judging the different rationing approaches and choosing among them. Alain Enthoven has devoted considerable attention to the information issue in recognition of this ([25], [26]). Nevertheless, it is a serious problem since consumers would need to know both the stated policies and the degree to which the plan follows them before informed choices could be made. For example, all members of a plan may prefer that organ transplants, sophisticated perinatal care for defective newborns, or expensive rehabilitation services be available. The plan may make these part of its offered treatments, but quietly ensure that very few patients are considered "medically suitable" for them. The patients' own families may not be aware that beneficial care has been denied, let alone the rest of the plan membership.

Also, regulatory mechanisms may not succeed in channeling competition among plans along socially desirable lines. In particular, there are strong financial incentives to control the composition of membership in a plan to avoid persons who use services intensively — to avoid enrolling them in the first place and to induce them to disenroll when their high utilization becomes apparent. These actions can be taken in subtle ways which are hard to control by regulation or market pressure. HMOs can use the geographical location of their facilities, waiting times, the quality level of particular departments, and a host of other variables to influence the composition of their patient populations. For example, if young families are profitable and older men entering the heart attack years are not, the HMO could let it be known that their pediatricians are first-rate but their cardiologists are only average [43].

Though neither NHS nor CCHP is perfect, both do set a level of care available to all that is a satisfactory first approximation to adequacy. In both, the process is dynamic and layers of decision making are involved. In both, there is a complex combination of political, administrative, and individual decision-making processes, although the balance of these is different in the two systems.

V. CLINICAL DECISION MAKING IN A RATIONING SYSTEM

In both the NHS and the CCHP, and in any likely alternative, doctors play a key role in setting the necessary limits on the care patients receive. In the case of the British National Health Service there are limits on resources and some explicit rationing rules. Aaron and Schwartz, however, explain various ways that doctors are able to circumvent these restrictions. These authors rightly stress the importance of physician cooperation with the process of limit-setting in the making of clinical decisions as a key factor in making the system work ([1], p. 103).

It is important to emphasize that this does not mean that doctors are performing cost-benefit analyses at the bedside. What actually happens in NHS or an HMO is an evolution of the standard of care into a cost-conscious one. Although doctors rail against "cookbook medicine" and emphasize the need for individualized treatment, there actually is a standard way of handling most medical conditions. If a patient sues for malpractice, another doctor can go into court and testify as to whether or not the accused doctor was practicing an acceptable standard of medicine. In a national health service, and to a lesser extent in an HMO, a resource-saving standard of care develops. The uncertainty in medicine about what really works and the variability in medical practice associated with it facilitates this development. In the area in which benefits are uncertain, doctors can permit other factors to influence their decisions and still satisfy their ethical beliefs. In these circumstances, they would not be denying *beneficial* care for reasons of cost. However, Aaron and Schwartz found that calling dialysis "not beneficial" for an individual was sometimes a way of rationing without acknowledging it ([1], p. 104).

Many find such an approach disturbingly dishonest. But if there is a societal consensus that cost is a legitimate factor in the receipt of care, and if it is not carried too far, a case can be made in its favor. There is an important sense in which something can be beneficial, but not worth having. If a fair process is employed to decide when benefits are not worth their costs, it is not unethical to say the care is not important or to deprive a patient of a costly treatment. If doctors were to be troubled by denying care to certain patients, their hesitation should be taken as an indication that the standard of care in fact might not be adequate. This is particularly likely to be the case in the United States which can afford a generous standard of adequacy.

Nevertheless, when beneficial care is denied, a serious ethical problem does arise. What should the patient be told? Should he be told something could be done, but it is not worth doing? Should he be told that nothing can be done? Or, should he simply be told nothing?

How much information ought a doctor in a rationing system provide about treatment options that are not available within the system? Bioethicists have devoted little attention to this. In recent years, the paternalism that formerly characterized the doctor-patient relationship has given way to an ideal of open communication and the exercise of patient autonomy. Yet the literature on patient-provider communication tends to focus on the patient's right to know rather than the practical problems this poses in a system in which options are limited.

Several possible models for doctor-patient communication in a rationing system can be suggested. They are not mutually exclusive and each can be found to some extent in the existing delivery system.

1. The patient is told of all beneficial options and allowed to choose, but finanical constraints are imposed which give incentives to make the correct choices. Traditionally, these choices are made through insurance policies with extensive cost-sharing and many exclusions. The provider is kept out of the process; all the incentives are on the consumer. Unfortunately, this conflicts with the consumer's goal in purchasing insurance protection against financial risk.

2. The patient is told of all beneficial options and makes his own choice, but the doctor exercises judgment about what the third party payor should pay. The patient can receive more care than the insurance company pays for (the doctor may even recommend more), but the patient would have to pay for it himself. The use of the physician as the company's advisor permits additional risk-spreading to be accomplished at reasonable cost by providing a more sensitive instrument for controlling utilization and the doctor is able to preserve his clinical autonomy.

Two difficulties arise with this approach. First, it may be feasible for major alternatives (as with surgery vs. medical management of a condition, an expensive drug vs. one that is cheaper but somewhat less effective), but where should the doctor draw a line? Need he explain that the hospital has chosen to use a lower quality of suture than the best available in the operating room? That a diagnostic test is being omitted that is very expensive, but has a one in a thousand chance of producing important information for his treatment? Such total communication surely requires more participation in the details of care than the patient wants. Yet, if the patient must consent to any limit on beneficial care, it would seem to be required.

A second difficulty arises if the patient is unlikely to be able to pay for the treatment. In some cases, might it not then be cruel and pointless to tell him of its existence? Should a Mississippi dirt farmer be told of an expensive cancer treatment available at Sloan-Kettering that has a one in a thousand chance of saving his life if there is only a one in a million chance that he could take advantage of it? Undoubtedly there are situations in which a patient would want to know about alternatives the doctor cannot provide, but must the doctor inform him of all alternatives and if not, where is the line to be drawn?

3. The patient consents in advance to the fact that he will not be offered all possible choices. For example, a person joining a health maintenance organization is told in advance that the HMO doctors will not inform him of all the options that are possible, but will omit some for reasons of cost. By joining, he consents to such a practice. In a national health service, it is understood that there are limits on choice.

Even if such an advance directive is ethically acceptable, there is probably still an obligation on the part of the doctor to inform a patient of options not provided when the doctor thinks that the patient might want to purchase the care outside the system or lobby for a change in the range of options provided. Again, the difficulty is knowing where to draw the line.

4. The medical profession defines the standard of practice in a way that limits the available options to those for which benefits bear a reasonable relation to costs (and to provide some guidance as to when exceptions to the rules are appropriate). An example would be consensus development conferences which allow cost to be a factor in the final recommendations. Patients then only are informed of those options which are part of standard medical practice. The NIH has used such conferences in recent years to influence medical practice with

respect to Caesarean deliveries, prenatal diagnosis, treatment of travellers' diarrhea, and other issues.

VI. CONCLUSION

A practical framework for health policy based on guaranteeing universal access to an adequate level of care is appealing. However, any system which delivers an adequate level of care at reasonable cost will require limits on beneficial care, and if these limits are to be imposed equitably, the cooperation of physicians is essential. An important consequence of this point is that the ethical duties of doctors in setting limits on patient care and in informing patients about the nature of such limits need to be clarified. An ethic which requires doctors to provide all beneficial care *whatever the cost* and to inform the patient in full about any choices that could conceivably affect the benefits he could receive has the charm of simplicity. However, it is incompatible with designing a practical system for delivering adequate care.

George Washington University
Washington, D.C.

NOTES

[1] In this approach, "ensuring equitable access to care" means everyone should be able to obtain an "adequate level" or "decent minimum" of care, the exact amount depending only on health state, without having to bear an "excessive burden" in financial cost, travel, and waiting time. People who want more than this level of care should be able to purchase it at its unsubsidized supply cost. The precise content of the adequate level and the definition of excessive burden are value judgments depending on society's trade-offs among different kinds of health care and between health care and other commodities. In this concept of equity, therefore, health care is considered differently from other goods in that it ensures that people can get care recognized as important, whether or not they can pay; yet, it requires neither a commitment of resources so open-ended that it jeopardizes the pursuit of other social goals, nor restrictions on the amounts people can spend on health care out of their own resources. The policy literature on the basic minimum or adequate level concept is extensive, see ([2], [25], [26], [36], [37], [40], [51], [52], [53]).

[2] Nevertheless, strong philosophical arguments can be and have been advanced in support of the decent minimum. These arguments are utilitarian, contractarian, or libertarian in nature. As such, they employ the premises of these ethical theories to examine the basis for believing that health care is special, the existence of a moral right to health care or alternatively, a societal obligation to provide it, and the extent of any

such right or obligation (that is, the content of the decent minimum). See, for example, ([9], [11], [13], [15], [16], [18], [19], [22], [24], [28], [32], [33], [34], [44], [45], [49], [57]).

[3] "Standard of care" is used in the sense that it is used in discussions of physician practice and in malpractice law — the compendium of routine ways of handling medical problems [48].

[4] The more concerned people are about the size of government expenditure, the more unwilling they are to believe that serious deficiencies in access to care exist. The reaction of the commissioners to a hearing held by the President's Commission is an example of this. At this hearing, individuals testified about their own problems in obtaining health care. Each case was carefully selected to illustrate a structural defect in the system, rather than a failure of the system to operate as designed. In other words, by logic, there was a large class of individuals who must have experienced similar difficulties. The commissioners who were particularly reluctant to enlarge the scope of government responsibility in health care were most vehement in objecting to this hearing on the grounds that it was "anecdotal". They also were the most reluctant to accept the standard statistical evidence presented to them on differences in access to care. A similar phenomenon can be seen with respect to hunger in the United States.

BIBLIOGRAPHY

[1] Aaron, H. J. and W. B. Schwartz: 1984, *The Painful Prescription: Rationing Hospital Care*, The Brookings Institution, Washington, D.C.

[2] Abel-Smith, B.: 1978, 'Minimum Adequate Levels of Personal Health Care: History and Justification', *Milbank Memorial Fund Quarterly/Health and Society 56*, 7—21.

[3] Aday, L., *et al.*: 1980, *Health Care in the U.S.: Equitable for Whom?*, Sage Publications, Inc., Beverly Hills, California.

[4] Agich, G. J. and C. E. Begley: 1985, 'Some Problems With Pro-Competition Reforms', *Social Science and Medicine 21*, 623—630.

[5] Arras, J.: 1981, 'Health Care Vouchers and the Rhetoric of Equity', *Hastings Center Report 11*, 29—39.

[6] Arrow, K. J.: 1963, 'Uncertainty and the Welfare Economics of Medical Care', *American Economic Review 53*, 941—973.

[7] Baily, M. A.: 1984, 'Rationing and American Health Policy', *Journal of Health Politics, Policy and Law 9*, 489—501.

[8] Barry, B.: 1965, *Political Argument*, Routledge and Kegan Paul, London.

[9] Beauchamp, T. L. and R. Faden: 1979, 'The Right to Health and the Right to Health Care', *The Journal of Medicine and Philosophy 4*, 118—131.

[10] Boulding, K.: 1966, 'The Concept of Need for Health Services', *Milbank Memorial Fund Quarterly 44*, 202—228.

[11] Branson, R. and R. M. Veatch: 1976, *Ethics and Health Policy*, Ballinger Publishing Co., Cambridge, Massachusetts.

[12] Brock, D. W.: 1983, 'Distribution of Health Care and Individual Liberty', in President's Commission for the Study of Ethical Problems in Medicine, *Securing Access to Health Care, Volume Two: Appendices: Sociocultural and Philosophical Studies*, U.S. Government Printing Office, Washington, D.C., pp. 239—263.

[13] Buchanan, A. E.: 1981, 'Justice: A Philosophical Review', in E. E. Shelp (ed.), *Justice and Health Care*, D. Reidel Publishing Co., Dordrecht, Holland, pp. 3—21.

[14] Buchanan, A. E.: 1982, 'Philosophical Foundations of Beneficence', in E. E. Shelp (ed.), *Beneficence and Health Care*, D. Reidel Publishing Co., Dordrecht, Holland, pp. 33—62.

[15] Buchanan, A. E.: 1983, 'The Right to a Decent Minimum of Health Care' in President's Commission for the Study of Ethical Problems in Medicine, *Securing Access to Health Care, Volume Two: Appendices: Sociocultural and Philosophical Studies*, pp. 207—238.

[16] Childress, J. F.: 1979, 'A Right to Health Care?' *The Journal of Medicine and Philosophy 4*, 132—147.

[17] Childress, J. F.: 1981, 'Priorities in the Allocation of Health Care Resources', in E. E. Shelp (ed.), *Justice and Health Care*, D. Reidel Publishing Co., Dordrecht, Holland, pp. 139—150.

[18] Daniels, N.: 1979, 'Rights to Health Care and Distributive Justice: Programmatic Worries', *The Journal of Medicine and Philosophy 4*, 174—191.

[19] Daniels, N.: 1981, 'Cost-Effectiveness and Patient Welfare', in M. D. Basson (ed.), *Rights and Responsibilities in Modern Medicine*, Alan R. Liss, Inc., New York, pp. 159—170.

[20] Daniels, N.: 1981, 'Health Care Needs and Distributive Justice', *Philosophy and Public Affairs 10*, 146—179.

[21] Daniels, N.: 1982, 'Equity of Access to Health Care: Some Conceptual and Ethical Issues', *Milbank Memorial Fund Quarterly/Health and Society 60*, 51—81.

[22] Daniels, N.: 1983, 'A Reply to Some Stern Criticisms and a Remark on Health Care Rights', *The Journal of Medicine and Philosophy 8*, 363—372.

[23] Engelhardt, H. T., Jr.: 1979, 'Rights to Health Care: A Critical Approach', *The Journal of Medicine and Philosophy 4*, 113—117.

[24] Engelhardt, H. T., Jr.: 1981, 'Health Care Allocations: Responses to the Unjust, the Unfortunate, and the Undesirable', in E. E. Shelp (ed.), *Justice and Health Care*, D. Reidel Publishing Company, Dordrecht, Holland, pp. 121—137.

[25] Enthoven, A.: 1978, 'Consumer Choice Health Plan', *New England Journal of Medicine 298*, 709—720.

[26] Enthoven, A.: 1980, *Health Plan: The Only Practical Solution to the Soaring Cost of Medical Care*, Addison-Wesley, Reading, Masschusetts.

[27] Fried, C.: 1975, 'Rights and Health Care Beyond Equity and Efficiency', *New England Journal of Medicine 293*, 241—245.

[28] Fried, C.: 1983, 'An Analysis of "Equality" and "Rights" in Medical Care', in J. Arras & R. Hung (eds.), *Ethical Issues in Modern Medicine*, 2nd edition, Mayfield Publishing Co., Palo Alto, California, pp. 490—496.

[29] Fuchs, V.: 1974, *Who Shall Live?*, Basic Books, New York.

[30] Fuchs, V.: 1984, 'The "Rationing" of Medical Care', *New England Journal of Medicine 311*, 1572—1573.

[31] Gauthier, D.: 1983, 'Unequal Need: A Problem of Equity in Access to Health Care', in President's Commission for the Study of Ethical Problems in Medicine

Securing Access to Health Care, Volume Two: Appendices; Sociocultural and Philosophical Studies, pp. 179—205.

[32] Gibbard, A.: 1982, 'The Prospective Pareto Principle & Equity of Access to Health Care', *Milbank Memorial Fund Quarterly/Health and Society 60*, 399—428.

[33] Green, R. M.: 1976, 'Health Care and Justice in Contract Theory Perspective', in R. M. Veatch & R. Branson (eds.), *Ethics & Public Policy*, Ballinger Publishing Co., Cambridge, Massachusetts, pp. 111—126.

[34] Green, R. M.: 1983, 'The Priority of Health Care', *The Journal of Medicine and Philosophy 8*, 373—380.

[35] Gutman, A.: 1981, 'For and Against Equal Access to Health Care', *Milbank Memorial Fund Quarterly/Health and Society 59*, 542—560.

[36] Havighurst, C. C.: 1977, 'Health Care Cost-Containment Regulation: Prospects and an Alternative', *American Journal of Law & Medicine 3*, 309—322.

[37] Havighurst, C. C.: 1981, 'Competition in Health Services: Overview, Issues and Answers', *Vanderbilt Law Review 34*, 1117—1158.

[38] Hirschman, A. O.: 1970, *Exit, Voice and Loyalty: Response to Decline in Firms, Organizations and States*, Harvard University Press, Cambridge, Massachusetts.

[39] Hirschman, A. O.: 1980, 'Exit, Voice, and Loyalty: Further Reflections and a Survey of Recent Contributions', *Milbank Memorial Fund Quarterly/Health and Society 58*, 430—453.

[40] Jonas, S., *et al.*: 1981, *Health Care Delivery in the United States*, 2nd ed., Springer Publishing Co., New York.

[41] Levinsky, N. G.: 1984, 'The Doctor's Master', *New England Journal of Medicine 311*, 1573—1575.

[42] Loewy, E. H.: 1980, 'Cost Should Not Be a Factor in Medical Care', *New England Journal of Medicine 302*, 697.

[43] Luft, H.: 1982, 'Health Maintenance Organizations and the Rationing of Medical Care', *Milbank Memorial Fund Quarterly Health and Society 60*, 268—306.

[44] McCullough, L. B.: 1979, 'Rights, Health Care, and Public Policy', *The Journal of Medicine and Philosophy 4*, 204—215.

[45] McCullough, L. B.: 1981, 'Justice and Health Care: Historical Perspectives and Precedents', in E. E. Shelp (ed.), *Justice and Health Care*, D. Reidel Publishing Co., Dordrecht, Holland, pp. 51—71.

[46] Mechanic, D.: 1976, *The Growth of Medicine: An Inquiry into the Dynamics of Patient Behavior and the Organization of Medical Care*, Wiley-Interscience, New York.

[47] Mechanic, D.: 1979, *Future Issues in Health Care: Social Policy and The Rationing of Medical Services*, The Free Press, MacMillan Publishing Co., New York.

[48] Morreim, E. H.: 1985, 'The MD and the DRG', *Hastings Center Report*, 30—38.

[49] Moskop, J. C.: 1983, 'Rawlsian Justice and a Human Right to Health Care', *The Journal of Medicine and Philosophy 8*, 329—338.

[50] Pellegrino, E. D.: 'Medical Morality & Medical Economics', *Hastings Center Report*, 8—12.

[51] President's Commission for the Study of Ethical Problems in Medicine and

Biomedical and Behavioral Research: 1983, *Securing Access to Health Care, Volume One: Report*, U.S. Government Printing Office, Washington, D.C.

[52] Rosenthal, G. and D. M. Fox: 1978, 'A Right to What?: Toward Adequate Minimum Standards for Personal Health Services', *Milbank Memorial Fund Quarterly/Health and Society 56*, 1—6.

[53] Schelling, T. C.: 1979, 'Standards for Adequate Minimum Personal Health Services', *Milbank Memorial Fund Quarterly Health and Society 57*, 212—233.

[54] Siegler, M.: 1979, 'A Right to Health Care: Ambiguity, Professional Responsibility, and Patient Liberty', *The Journal of Medicine and Philosophy 4*, 148—157.

[55] Siegler, M.: 1980, 'A Physician's Perspective on a Right to Health Care', *Journal of the American Medical Association 10*, 1591—96.

[56] Stern, L.: 1983, 'Opportunity and Health Care: Criticisms and Suggestions', *The Journal of Medicine and Philosophy 8*, 339—362.

[57] Veatch, R. M.: 1979, 'Just Social Institutions and the Right to Health Care', *The Journal of Medicine and Philosophy 4*, 170—173.

[58] Willard, L. D.: 1982, 'Needs & Medicine', *The Journal of Medicine and Philosophy 7*, 259—274.

[59] Williams, A.: 1978, 'Need: An Economic Exegesis', in A. J. Culyer and K. G. Wright (eds.), *Economic Aspects of Health Services*, Robertson and Co., London.

DAVID D. FRIEDMAN

COMMENTS ON "RATIONING MEDICAL CARE: PROCESSES FOR DEFINING ADEQUACY"

Mary Ann Baily begins her paper by setting out, in general terms, the idea of a "decent minimum approach" to providing health care. She goes on to argue for defining the "decent minimum" as the level that should be (but not the level that is) chosen by the middle class, and to discuss reasons why it should and ways in which it could be made available to everyone.

Baily's paper is built around a normative proposition — that everyone should get a "decent minimum" of health care. The author does not herself provide a defense of this proposition. Instead, she cites a variety of other writers who have defended it, and devotes her paper to discussing how the objective might be achieved.

While the paper takes the form of a positive discussion of how certain goals are to be achieved, goals which the author observes that other people are in favor of, I believe that it can and should be discussed in both normative and positive terms. A large part of the paper is devoted to discussions of normative issues such as what a "decent minimum" means and all of the paper is motivated by a normative issue. Further, the paper itself is shot through with normative language. Phrases such as "decent minimum", "it would be a moral outrage if. . . ," "adequacy," and the like cannot be described as value-free descriptions of alternative objectives. Hence, in the first part of this commentary, I will deal with normative issues that the paper raises. In the second part, I will go on to consider the positive issues.

I. PHILOSOPHY

Baily's essential normative proposition is that everyone should be provided with a "decent minimum" of health care — and hence that if some individuals are unable to provide that decent minimum for themselves, it should be provided for them by others. Underlying most arguments in favor of this proposition is the claim that health care is in some sense a special good — essentially different from most or all of

G. J. Agich and C. E. Begley (eds.), The Price of Health, 185–197.

the other things humans desire. While I cannot — short of writing a review article on Baily's entire bibliography — deal with all such arguments, I will try to deal with the most common.

It is often claimed that health care is special because it is a "need" whereas most other things are only "wants". This seems to reduce either to the assertion that health care is a need because it is necessary for life, and life is infinitely valuable relative to other desirable things, or that it is a need because the kind and amount of health care which an individual requires is a matter of objective fact, to be determined by experts, rather than a matter of preferences.

The claim that life is infinitely valuable, while rhetorically satisfying, is difficult to defend, whether as a proposition about how people do act or about how they should act. One piece of evidence that might be offered for the proposition that individuals act as if the value of their own lives were infinite is the observation that few people would be willing to sell their lives in exchange for a cash payment, even a very large one. But that is evidence, not that life is infinitely valuable, but that money is of no use to a corpse. We do observe people giving up their lives in situations where their objective — the price for which the life is being sold — does not depend for its value on their remaining alive. One obvious example is the soldier going "over the top" in the First World War.

We rarely find someone sacrificing his entire life for some valued objective, but we frequently observe people sacrificing small parts of their lives in exchange for quite minor rewards. There are many individuals who smoke even though they believe that smoking will reduce their life expectancy. People ride motorcycles, jump out of airplanes, drive dynamite trucks, and in many other respects accept a decreased life expectancy in exchange for either payment or pleasure. What distinguishes these cases from the individual selling (or refusing to sell) his life for a million dollars is that if one "sells" a one percent chance of dying, there remains a 99 percent chance of remaining alive to spend the money one has been paid.

All of this is evidence that people do not regard their own lives as infinitely valuable. A possible response is that even if they do not, they should. If one tries to imagine this principle embodied in a real society, however, it is one which few would find attractive. To assert that life is infinitely valuable is to assert that all finite values should be sacrificed to it, hence, a society built on that principle would be one in which

100 percent of national income was spent on maximizing life expectancy. No entertainment – except to the extent that it increased life expectancy. No avoidable risk taking. All food prepared solely on the basis of nutrition. Athletics forbidden. Calisthenics, which provide the same benefits without the risk of accident, compulsory for all. Childbirth either forbidden (if one is trying to maximize per-capita life expectancy) or compulsory (if one is trying to maximize total life expectancy).

Finally, for any who remain convinced that medical care is special because of its connection to life, it is worth noting that the same connection exists for a very wide range of other things. Improved medical care has played an important role in the increase of life expectancy over the past fifty years. But most of the increase in life expectancy over the past two centuries occurred earlier – during a time when it was an open question whether medical treatment increased or decreased a patient's chances of recovery. One of the main causes of increased life expectancy appears to have been the enormous increase in real income over that period – expressed as shorter and easier hours of work, better nutrition, better housing, better clothing, and a wide variety of other goods. If we were somehow to construct a society in which the sole objective was to maximize life expectancy, it would be very different from the society we now live in – but it is not certain that it would spend more of its resources on medical care.

If one reason to regard medical care as a need is the claim that life is infinitely valuable, another is the claim that medical needs are a matter for objective decision by experts, unlike wants, which depend on the subjective preference of the consumer. But whether a patient should have a treatment does not depend entirely on the effect the treatment will have on the patient's health. It also depends on the value of that health to the patient, and on the value to the patient, or whoever else is paying for the treatment, of the goods that must be given up in order to pay for it. An expensive operation which is obviously appropriate to a professional athlete may be just as obviously inappropriate to an elderly and sedentary gentleman, who is happy to accept some minor cost in decreased agility in order to avoid a week in the hospital. And a treatment providing a minor benefit may be obviously desirable to a rich man, for whom the cost is not eating caviar for a week, and obviously undesirable for a poor man, for whom the cost is not eating for a week.

This brings me to an issue which underlies much of the discussion of health care – the relation between income and medical care. My purpose here is not to argue that the existing distribution of income is just, or even that any inequality in income is just. My purpose is simply to argue that there is nothing special about medical care. If, on grounds of justice or utility, the poor should have more money, there is no reason why they should be given it only on the condition that they spend it on health care.

Putting the question in this way may make more plausible my point about the rich man and the poor man. Suppose, for a moment, we take it as given that we are going to have a society in which some people are richer than others. Suppose we also take it as given that some resources are going to be taken from those with higher incomes and given to those with lower incomes. The assertion that the poor have a right to a "decent minimum" of health care is then equivalent to insisting that some of the transfer must be in the form of medical care – whether given directly or paid for via a voucher or an insurance policy.

Imagine, then, that we have decided that a particular poor person will receive a transfer of three thousand dollars a year. The issue is whether he should receive it in cash or in the form of medical insurance sufficient to provide a "decent minimum" of health care. If he receives it in cash, he can choose to spend the money either on medical insurance or on whatever else he prefers – perhaps food. If the standard of care in question is that appropriate to the middle class, as Baily suggests, then to insist on giving the subsidy in the form of medical insurance sufficient to provide a decent minimum, instead of giving it as cash and letting the recipient decide whether or not to spend it on medical insurance, is equivalent to telling the poor man that he must give up his dinner to pay for some item of medical care – on the grounds that it is an item for which a member of the middle class would be willing to give up his dessert. This seems a hard position to defend. Yet, it is the position implied by the idea that medical care is a special good, that everyone is entitled to a "decent minimum," and that a "decent minimum" should be defined as the level of consumption appropriate to the middle class.

I have been dealing with arguments for the "decent minimum" based on the idea that health care is a need. There are, of course, other (although not, in my opinion, better) arguments in favor of the same conclusion, some of which I have discussed elsewhere [2]. Rather than pursue those arguments here, I would prefer to raise a different, but not unrelated issue.

It may seem to many readers that I have demonstrated — or claimed to demonstrate — too much. What I have been arguing is not merely that the idea that everyone should be guaranteed a "decent minimum" of health care is wrong, but that it is indefensible. If this is so, why do so many people find it convincing? One point Baily makes which is, I think, correct, is that most people are reluctant to admit — to themselves as well as to others — that the level of health care which people receive depends on their income, and that many poor people receive a level of care that the average American would regard as entirely unsatisfactory for himself. And while I think she implies a somewhat greater philosophical consensus in favor of a "decent minimum" than actually exists — I am still waiting to meet the libertarian supporters of the position whose existence is, if not asserted, at least suggested in one of her notes — she is surely also correct in believing that many philosophers support the idea.

In trying to understand why the belief that medical care has some special priority is so widely held, I find introspection a useful tool. I imagine a case where there is a choice between life for one person and some minor pleasure for a very large number of people. To make the example concrete, I imagine that there is some individual who, to preserve his life, requires a medical treatment which will cost fifty million dollars — and is not going to get it. I further imagine that there are a hundred million people in the United States who use mint flavored toothpaste, and that the cost, per person, of putting mint flavor in toothpaste is fifty cents a year. It seems reasonable to describe this as a situation in which one man is losing his life in order that other people can have mint flavor in their toothpaste. Is this in some sense a correct outcome, or should the hundred million people be required (or choose) to forego mint flavor for a year in order to pay for the operation?

I think the natural response of most people — including most libertarians and including myself — is that this outcome is not correct, that life is more important than mint flavored toothpaste, and that a good society should express that fact in its outcomes.[1] I also think that that natural response is wrong. Where is the error?

The error, it seems to me, is in my inability to intuit large numbers. In thinking about such a situation, I try to compare two values — the value of life to one person and the value of a minor pleasure to a hundred million. Since I cannot imagine a hundred million people — I have no clear intuition for what a number that large means — I end up imagining the situation as if it were a tradeoff between life for one person and mint flavor for five or ten or (if I have an extraordinarily good imagination) twenty. In that case, the answer is obvious — one life is worth much more than mint flavor for twenty people.

I try to help my intuition by switching to a probabilistic version of the same problem. Suppose I do not know, in advance, who it is who will need the operation — merely that it is one of the people who enjoys mint toothpaste. Further suppose that I am such a person. I am then deciding whether I would be willing to give up mint flavor for a year in order to eliminate one chance in a hundred million of death. Now the answer is very much less clear. My intuition is no better able to deal with a probability of one in a hundred million than with a population of a hundred million. Whatever the right answer may be, when I think of it as a decision I make for myself it no longer seems absurd or monstrous to choose a minor pleasure in exchange for (a very small chance of losing) life.

As a further prop to my intuition, I try converting a small probability of losing a large amount of life into a certainty of losing a small amount. While there is no a priori reason why we must be risk neutral with regard to life, it seems as though the most natural comparison is one in which the expected value of the loss is held constant. That means that one chance in a hundred million of losing all of (the rest of) my life is equivalent to a certainty of losing one hundred millionth of it — or about thirteen seconds. That does not seem like a wholly unreasonable price to pay for the minor pleasure of mint flavored toothpaste.

My conclusion is that the idea of providing a "decent minimum" of health care is indeed attractive to many people, but that one important reason for its attractiveness is an error in our moral intuition — our inability to intuit large numbers or small probabilities. Once one has eliminated that error, there may remain legitimate arguments for helping the poor and legitimate arguments for a variety of government interventions in the medical market, possibly including those that Baily suggests. What does not remain is any justification for the claim that the poor have a special claim to a particular level of medical care, above and beyond any general claim that they may have to resources, equality, and the like.

Baily might reply that even if the idea of guaranteeing a "decent minimum" is philosophically indefensible, many people still believe in it, and that her business as an economist is to tell people not what they should want but how to achieve what they do want. In the second part of my comments, I will discuss her paper from that standpoint. I will assume that the objective is to provide everyone with "a standard of care acceptable to the middle class" and that the two questions of interest are whether it can be done at a reasonable cost — more precisely, at a per-capita cost considerably lower than the current per-capita cost of medical care for the middle class — and if so how.

Before going on to do so, I would like to end my discussion of the philosophical bases for the "decent minimum approach" by making a point based on a sentence of Baily's: "For practical policy, the difficulty is not that the adequate level is characterized differently by the different approaches, but that these approaches do not provide a concrete characterization of adequacy at all". When one encounters a set of philosophical "arguments" all of which lead to the same "principle" and none of which provides any way of telling whether that principle is being met, the proper response is profound scepticism — and a strong suspicion that what one has actually met is a collection of campaign speeches in fancy dress.

II. ECONOMICS

Doing it Better This Time

Baily makes it clear that she is not proposing to extend to the poor the standard of care currently received by the middle class, which she describes as "wasteful and inefficient", but rather the standard that should be received by the middle class — "the standard such a person would choose if the costs were reflected in actual health insurance premiums and taxes". This raises a problem which she never faces. The existing situation is not an accident. It is a datum — a piece of evidence about how a particular system works. As such, it cannot simply be assumed away. Until one has shown why our system of health care delivery is wasteful and inefficient, and how it is possible to make it radically less so, one must include the observed faults of our present system in any future projection.

It is least arguable that the wasteful and inefficient nature of our system is largely due to the degree to which health care, like schooling, another good whose provision is generally regarded as wasteful and inefficient, is controlled or provided by political mechanisms. An important element in this is the de facto grant of power to the medical profession to control its membership, via licensing and similar restrictions. That power has been used as one might expect — to restrict the number of practitioners, thus raising their wages, to encourage insurance institutions to develop in ways which maximize the demand for medical care, and to discourage the development of institutions intermediate between doctors and patients, such as HMOs, that would help the customer to make a rational tradeoff between costs and benefits [4].

Baily comments that in the United States, "many physicians feel it is unethical to consider cost in their clinical decisions". She does not discuss the obvious economic consequences of that "ethical" position. Some who comment on this subject appear to regard the unwillingness of medical professionals to restrict care on the basis of cost as evidence of their generous and charitable nature. It is worth at least noting that this particular kind of generosity imposes no cost on those who adopt it, since they are not the ones paying the bill.[2] Not only does it impose no cost on them as individuals, it is very much to their benefit as a profession. A policy of "cost is no object", implemented by individual physicians and embedded in the structure of traditional health insurance, results in an increased demand for medical services, hence more jobs and higher wages for medical professionals. Praising the medical profession for its unwillingness to ration care on the basis of cost is like praising General Dynamics and Hughes Aircraft for their patriotic dedication to the nation's defense.

Doubtless, most medical professionals believe their own rhetoric — just as defense contractors believe in the importance of defense, teachers that it is essential to spend more money on schooling, car manufacturers that the importation of Japanese automobiles threatens the economic health of the nation, and economists that the United States should employ more economists. The issue is not one of moral turpitude. In a complicated society, most of us are able, without too much effort, to persuade ourselves of the identity between our private interest and the welfare of mankind.

It is also worth noting that the effects of government regulation of the medical industry differ very little from the effects of regulation of other industries. According to the folklore of regulation, it is imposed upon unwilling industries in order to prevent them from mistreating their customers. It is now fairly well established that the typical pattern is just the opposite.[3] Regulation is usually established at the request of the regulated industry, it is usually controlled, in large part, by the regulated industry, and it usually functions to prevent competition and thus strengthen the position of the industry vis à vis its customers.

This is not surprising. In our political system, perhaps in any political system, concentrated and well organized interest groups have more power than dispersed and poorly organized groups. Typically, although not inevitably, an industry is very much more concentrated and better organized than its customers. Hence, typically, government involvement serves to benefit the industry at the expense of its customers.[4] There is no obvious reason why that logic would apply any less to the medical industry than to others — whether under present institutions or under the various future institutions that Baily discusses.

I have gone into this point at such length because an essential element of Baily's argument is the implicit claim that, in implementing a "decent minimum" standard, one can somehow wave a magic wand and eliminate the wasteful and inefficient features of present-day medicine. But one of the conditions which Baily insists on, for all of the alternatives she considers, is that "the process must be linked to the political process". If, as I have just argued and as others have argued before, the linkage between the provision of medicine and the political process is a principal cause of the inefficiency of the present system, then we may expect any system similarly linked to be similarly inefficient.

It is of some interest to apply these remarks to the two systems that Baily discusses — the British National Health Service and the Consumer Choice Health Plan (CCHP). For the former, we have some empirical evidence, since it exists; for the latter, we must depend on theory.

Baily's main conclusions about the National Health Service are that the system is complicated and that it "seems to work". The evidence for the latter proposition is that, until recently, most Britons, having already paid for the NHS through their taxes, chose to take what they had paid for rather than paying a second time for private care. By that criterion, American public schools work too.

One point which Baily does not make, save very indirectly, is that if the criterion of a decent minimum is that the system provides what middle class people believe that they themselves should get, then the NHS does not provide a decent minimum. The evidence is the case of kidney dialysis. If the British population believed that denying dialysis to the elderly was consistent with a decent minimum, the NHS would have no need to pretend that it was merely avoiding dialysis where it was "contraindicated".

In discussing the CCHP, the first thing to note is that the plan, as described by Baily, is "market oriented" only relative to alternatives such as the NHS; it proposes a degree of government involvement in the health industry considerably greater than is typical of other industries in our society. The second thing to note is that, throughout her discussion, Baily treats the government as if it were a benevolent philosopher king, standing outside the system and adjusting it, where necessary, to compensate for the errors (or selfishness) of the participants.

As my previous discussion suggests, the question she should be asking is not: How can the government eliminate this or that imperfection in the outcome of the market?, but rather: Given what we know about the economics of regulation, what are the ways in which the government can be expected to intervene in the system to benefit some groups at the expense of others, and how seriously will such intervention undermine the working of the system? Since she is describing a system in which the government has more, not less, power than it does now, it seems reasonable to expect that the degree of intervention, and the resulting damage, will be greater under the CCHP than it is under present institutions.

Laundry List Economics

The obvious approach for an economist who wishes to analyze the desirability of a set of institutions is to specify an objective and then see how well the institutions can be expected to achieve it. Baily follows a different, and to me rather puzzling, procedure. She states her objective: "translating the decent minimum approach into practical policy". She then argues that a system which achieved that objective would have certain characteristics — it must be an ongoing process capable of incorporating changing information; it must allow individual providers to ration; it must be linked to the political process; and so on. She then shows, or at least argues, that there are two (and presumably many more) alternatives which have these characteristics. Finally, she concludes that she has demonstrated "that a practical system can be developed along the lines described".

This appears to involve a serious logical error. Even if it is true that an adequate system has a certain laundry list of characteristics, it does not follow that anything which has those characteristics is an adequate system. Man is a featherless biped, but a plucked chicken is not a man.

In the final sentence of her discussion of the two alternatives, Baily writes: "In both, there is a complex combination of political, administrative, and individual decision making processes . . . ." That is no doubt true. It is also true of the government of the United States, the government of the U.S.S.R., the Mafia, a concentration camp, and almost any other complicated human enterprise. The relevant question is not whether there is a complex combination, but whether that particular complex combination correctly balances the relevant costs and benefits.

For an economist whose objective is economic efficiency, correctly balancing costs and benefits means providing services when their value is greater than their cost and not providing them when it is less. For a utilitarian philosopher, it means taking all actions, and only those actions, which result in a net gain of utility. For other individuals, with other objectives, the correct balance will be a different one. Even if all were to agree that the ideal system was "dynamic", and involved "layers of decision making", and met every other criterion on Baily's list, the fact that a system met those criteria would be very poor evidence that it was achieving those objectives. A system designed to *minimize* efficiency, or utility, or the welfare of the least well off would probably also be "dynamic" and "many layered" and "a complex combination of political, administrative, and individual decision making processes . . . ."

III. CONCLUSION

Many Ann Baily, in her paper on rationing medical care, makes a number of interesting points. Nonetheless, the paper seems to me to be fatally flawed both on the normative level, as philosophy, and on the positive level, as economics. On the normative level, the paper works out the implications of a vaguely defined ethical proposition which, I believe, cannot be defended. On the positive side, the author tries to demonstrate that there exist a variety of alternative systems of health care which would both avoid some of the faults of our present system and provide a "decent minimum" to all. She neither analyzes the sources of the faults in our present system nor shows that they would be avoided in her proposed system. While she describes two alternative systems as providing a "decent minimum", she provides no substantial arguments to show that is how they would in fact work.[5]

A.B. Freeman School of Business
Tulane University
New Orleans, Louisiana

NOTES

[1] This does not mean that libertarians would endorse governmental transfers but rather that, believing a libertarian society would be a decent society, they would expect the problem to be taken care of somehow — perhaps by charity or private insurance.
[2] There are, of course, exceptions, such as the nursing orders. As a general rule, however, medical professionals do not donate their services for frec to their customers.
[3] In the case of the medical profession, the argument goes back at least to Friedman and Kuznets ([3], pp. 8—20, 118—137). For a general discussion of the theory of regulation, the reader may wish to look at Posner [6]. A nice example of the contrast between the real and the mythical history of regulation is Olasky's recent article on utility regulation [5].
[4] The logic of this situation is discussed at greater length in [1] and [2]. There are exceptions. In some cases a highly dispersed industry might be regulated by a concentrated set of customers. In other cases, political pressures unrelated to the interests of the industry may, at least temporarily, produce results unattractive to the industry. Examples of the latter would include the effect of the Thalidomide case on drug regulation and Nader's effect on safety regulation. In both cases, it could be argued that the regulation was eventually used by the industry in its own interest — to slow the introduction of new drugs, thus increasing the value of old ones, and to make it more difficult for foreign auto firms to compete in the American market.
[5] The distinction between discussions of government involvement that include government as part of the system being analyzed and those which treat government as a *deus ex machina* is explored at much greater length in [1] and [2].

BIBLIOGRAPHY

[1] Friedman, D.: 1986, *Price Theory: An Intermediate Text*, South-Western Publishing, Cincinnati.
[2] Friedman, D.: 1986, 'Should Medical Care be a Commodity?' in *Rights to Health Care*, D. Reidel Publishing Co., Dordrecht, Holland (in press).

[3] Friedman, M. and Kuznets, S.: 1945, *Income in Independent Professional Practice,* National Bureau of Economic Research, New York.
[4] Havighurst, C. C.: 1978, 'Professional Restraints on Innovation in Health Care Financing', *Duke Law Journal 1978,* 303—387.
[5] Olasky, M. N.: 1986, 'Hornswoggled', *Reason 17,* 29—33.
[6] Posner, R.: 1974, 'Theories of Economic Regulation', *Bell Journal* 5, 335—358.

GERALD R. WINSLOW

RATIONING AND PUBLICITY

I. INTRODUCTION

Should it not seem strange that so much attention is now being given to the *prospect* of rationing medical care in the United States? After all, if by "rationing" we suggest that access to medical care is limited in various ways, or that some people do not get the *best* care or even the *basic* care they need, then surely medical care always has been "rationed" in the United States ([11], [19], [32]). Among the fundamental reasons for such limits to care is one that is unlikely to change: The human capacity to develop new, often costly, and sometimes very scarce therapies outstrips the human capacity to make all such therapies available to everyone in need. So *some* method of restricting access must be a permanent feature of any system of providing medical care. Why, then, the present stir about rationing?

Full answers to this question are bound to be complicated. But the sort of person likely to be reading this essay is nearly certain to be acquainted with a long list of developments that have led to the recent talk about the inevitable rationing of medical care. Heading the list is concern about controlling runaway costs. These rising costs are attributed variously to increases in expensive new technologies, the loss of traditional market control of access because of a burgeoning system of third party payments, the greater needs of an increasingly aged population, and so forth. If we are to control costs, it is said with considerable frequency, we must face squarely the question of limiting access to care [1]. And to do this in some rational way, the emerging wisdom says, we must conduct careful cost/benefit or cost/effectiveness studies in order to ascertain which treatments are, in fact, *costworthy* [20].

It is not the purpose of this essay to challenge this emerging view. But I do want to reflect on one of the moral "costs" of discovering costworthiness and acting rationally on the information. It is the cost of losing innocence, as we are forced to face explicitly the fact that some of the lives that could be saved by medical care are simply not worth saving.[1]

G. J. Agich and C. E. Begley (eds.), The Price of Health, 199–215.

What is remarkable about our present situation, it seems, is *not* that we are moving from a time of no rationing to a time of much rationing. Indeed, from the standpoint of the distribution of medical care, it must seem ironic that so much attention is now focused on rationing just when more people are getting more care than ever before ([31], [29]). But what *is* remarkable is that the increasingly corporatized and rationalized structure of our health care system, with its growing capacity to ascertain the costworthiness of care, must lead away from the limits that were *implicit* in earlier days and toward rationing that is now *open to view.* The reality of rationing is likely to become all the more visible as the ideology of the "decent minimum" continues to develop strength and sophistication, thus making it obvious that a two-tiered (or, to be far more accurate, a *multi*-tiered) medical care system is not only a permanent feature of our society but also an officially sanctioned one.[2]

My question is: Should the rules for rationing lifesaving medical care be made public? Presumably, if rationing is becoming more obvious, then some people should know how it is being conducted. It does not follow, however, that the norms for rationing are widely understood by (or even accessible to) the general public. We may know, for example, that not all children with biliary atresia will become recipients of transplanted livers. But relatively few people in society know precisely how the selection process works. And when it does become clear just how one infant, a Jamie Fiske for example, has gained access to a transplantable liver, while others may have died for want of such a resource, an outburst of public concern about the equity of our present approaches to rationing may ensue ([12], [32]). So, for all the talk about explicit rationing, the rules for such rationing, it seems, are not much publicized. The question is: Should they be?

I want to answer this question affirmatively while, at the same time, acknowledging that such an answer is complicated, probably controversial, and maybe expensive.

Before proceeding, I should clarify three key terms. First, by the condition of "publicity" I simply mean that rules (in this case, the rules for rationing medical care) are made known to all the members of society who care to know them. Second, I am using "rules" rather broadly for all norms, either actual or proposed, that prescribe the way in which rationing should be conducted. Finally, by "rationing", I refer to any system of limiting access to medical care, especially life-saving medical care.[3] Again the usage is rather broad and could refer, for example, to systems with considerable centralized control as well as to free-market systems.

My focus, then, is primarily on what now are often called *micro-allocation* decisions; that is, which patients should be selected when potentially lifesaving medical resources are scarce? But the position that I am taking with regard to publicity should hold *a fortiori* for *macroallocation* decisions, or the decisions about which medical care programs society should give priority (or even what portion of society's resources should go to medical care). Indeed, judging from experience with rationing hemodialysis, there is reason to believe that publicizing the way in which microallocation decisions are made will heighten political interest in macroallocation decisions ([9], [27], [34]). And this is as it should be, or so I shall argue.

The rest of the essay is divided into three sections. First, I shall review very briefly some of the standard arguments for publicity. Next, I shall take up some of the objections to publicizing the rules for rationing that either have been or might be made. Finally, I shall attempt to answer these objections and offer a few reflections on how the condition of publicity might be attained in the practice of rationing medical care.

II. THE CASE FOR PUBLICITY

So ordinary and so little debated is the expectation of publicity for most of the rules that govern the social institutions of Western democracies that rehearsing the case for publicity might appear entirely unnecessary. Is it not an obvious requirement of participatory government that the citizenry be privy to the rules that are supposed to govern social interactions? And does not every middle school child already know that the rules must be announced and understood *prior* to the game?[4]

Still, there should be some merit in reviewing, even if briefly, some of the main philosophical arguments that have been offered in behalf of publicity.

1. *Publicity discourages the institution of inequitable rules that are incapable of universalization.* It is a widely accepted test of moral rules that they should be capable of being acted on by everyone ([3], pp. 195—200). Although publicity will guarantee neither universality nor equity, it can serve as an important aid in achieving these conditions by

making it obvious that certain rules cannot reasonably be applied or acted on by all members of society. This is part of what Immanuel Kant had in mind when he enunciated two "transcendental principles" of public law: (1) "All actions that affect the rights of other men are wrong if their maxim is not consistent with publicity." (2) "All maxims that *require* publicity (in order not to fail of their end) agree with politics and morality" ([13], pp. 135 and 139). If maxims must be kept secret in order to succeed, Kant argued, it is likely that they threaten the citizenry with injustice or unhappiness. Publicizing the rules helps to assure that the "rights of the public" will be protected.

2. *Publicity enhances the stability of cooperative morality by enlisting the support of society's members.* Kant also taught that publicity builds up public trust and serves cooperative morality by inclining people (and nations) to honor right rules and to identify with goals of cooperation. Publicity serves to maintain cooperative morality because people who understand the rules and see them to be in the service of social cooperation are more likely to follow the rules. Thus, it should be fairly obvious that the long-range stability of *cooperative* social institutions depends, in part, on the acceptance of *publicly* acknowledged rules.

3. *Publicity is necessary in order for moral rules to be taught.* Kurt Baier has argued most convincingly that the moral rules must be "completely universal and open" because "morality is meant to be taught to all members of the group in such a way that everyone can and ought to act in accordance with these rules" ([3], p. 195). The members of a society have a responsibility to teach moral rules to their children and to new members. In order to meet the condition of universal teachability, the moral rules obviously cannot be esoteric.

4. *Publicity aids the establishment of valid, mutual expectations.* One frequently acknowledged form of injustice is the abrogation of persons' legitimate expectations ([4], pp. 21—22). (Common examples include changing the rules in the middle of a game and enacting retroactive laws that contravene what people, through previous laws or long tradition, had come to count on.) Publicity is obviously not the only way to establish legitimate expectations; they may be created by custom, secret promises, or in many other ways. But publicizing the rules is one of the most secure ways of assuring that expectations are understood and met. In John Rawls's words: "The publicity of the rules of an institution insures that those engaged in it know what limitations on conduct to expect of one another and what kinds of actions are

permissible. There is a common basis for determining mutual expectations" ([25], p. 56).

5. *Publicity assists those who need to appeal injustices.* Knowledge of the rules not only establishes legitimate expectations, it also provides a basis for complaint if those expectations are unfairly abrogated. Without publicity it is difficult to imagine how people who believe that they have suffered injustices would be able to mount effective appeals. The ability to appeal would seem to be especially important when the decision involves potentially life-saving goods. Thus, the requirement of making available the basis for rationing medical care should be strong. This point was made a few years ago by the authors of a law journal article on the allocation of scarce life-saving medical resources:

> Due process generally requires decisionmakers to produce a record of their proceedings and an opinion stating the facts and reasons relied upon. The purpose is to enable individuals to ascertain that their rights have been properly protected, and to provide a possible basis for *recourse* [emphasis supplied] ([8], pp. 1748–1749; also [21]).

6. *Publicity promotes the criticism of rules that are in need of change.* This reason for publicity is particularly important in a democratic society whose citizens are supposed to be able to work for orderly change of social institutions. Clearly, such work is hindered if the rules governing the institutions are hidden from public view. Moreover, the quality of debate about social policies is likely to be low if the true nature of those policies is obscure.

I have not attempted to list the major arguments for publicity in any order of importance. And, without doubt, the list could be lengthened. But I do hope that I have given some sense of the reasons for considering publicity to be a highly significant condition for a truly cooperative, social morality. If there is one reason that undergirds all of those just cited, it is this: *Publicity is generally required in order for social institutions to manifest a commitment to respect persons as autonomous beings.* In this regard, publicity is akin to (though certainly not identical with) the requirement of informed consent to medical care. Without publicity, people will often be unable to evaluate their real alternatives. And, for people who need scarce life-saving medical care, obscurity of the rules that govern rationing decisions can hardly be accounted a matter of trivial concern.

For all the reasons just stated, I contend that the rules for rationing medical care should be subject, *prima facie*, to the requirement of publicity. Before exploring a few of the practical ramifications of this argument, it is important to consider, and, if possible, overcome some objections to such publicity.

III. OBJECTIONS TO PUBLICITY

In the history of moral philosophy, arguments for various kinds of "noble lies" are not unknown.[5] But, so far as I am aware, modern examples of explicit arguments for nonpublicity are rare.

One noteworthy exception is Henry Sidgwick's classic work, *The Methods of Ethics*. Sidgwick was concerned about how an "enlightened Utilitarian" should relate to conventional rules of morality in a society populated with a majority of people who do not share such enlightenment. So, for example, the utilitarian might believe that lying is productive of the greatest amount of good under certain circumstances. But teaching ordinary people, who are likely to be inclined to lie for all the wrong reasons, that it is sometimes permissible to lie could have disastrous effects on the moral strength of society. This danger should be reduced, according to Sidgwick, by limiting publicity; the utilitarian should not advocate publicly all aspects of his or her morality. And, in a remarkable passage, Sidgwick contends that the belief that not all rules should be made public should itself be kept from the public:

> [T]he utilitarian conclusion, carefully stated, would seem to be this; that the opinion that secrecy may render an action right which would not otherwise be so should itself be kept comparatively secret .... Or if this concealment be difficult to maintain, it may be desirable that Common Sense [i.e., the intuitive morality of ordinary people] should repudiate the doctrines which it is expedient to confine to an enlightened few. And thus a Utilitarian may reasonably desire, on Utilitarian principles, that some of his conclusions should be rejected by mankind generally .... ([30], p. 490).

Sidgwick's concern that unhappy results might follow full publicity is shared by some who have questioned public disclosure of the bases for rationing lifesaving goods. If, for example, people have to face the awful reality that some members of society will die due to scarcity made necessary by economic trade-offs, there may be a great deal of unhappiness. Would it not be better for a few people to bear the burden of this knowledge rather than to make the whole populace share in the miserable awareness?

The answer, according to Steven Rhoads, is "yes" [28]. Rhoads considers both micro and macroallocation decisions and their relationship to various approaches to valuing life. He concludes that we should sustain, so far as possible, the common belief in the pricelessness of each human life. This may be done, for example, by spending far more to save one identifiable person in need than it would be reasonable to spend for, say, new safety measures that might save many (as yet unknown) lives. But even as we seek to preserve the social conviction that each life is invaluable, key policy makers should use sophisticated information about the monetary value of life, especially in making macroallocation decisions. These policy decisions should be mostly hidden from public view; it is simply too demoralizing for society to acknowledge publicly that some lives are not worth saving. "As for candor," Rhoads concludes in a passage reminiscent of Sidgwick,

> I would tolerate a little dissembling in this area .... Admittedly, the absence of total candor leaves the way open for the periodic appearance of ambitious reporters and politicians who expose the Dr. Strangelove-like analysts at the heart of the bureaucracy. But this cost seems tolerable in an area where simple enlightenment is not to be expected or wished for ([28], pp. 305—306).

We should notice, of course, that Rhoads has broken one of Sidgwick's rules, namely, that the principle of non-publicity should itself be kept secret. But the irony of *publishing* an essay advocating the use of value of life analyses while at the same time contending that the public should not be burdened with such demoralizing thoughts, has not been lost on Rhoads. In a footnote that might have pleased Sidgwick, Rhoads reassures himself and us that

> this chapter and book are not inconsistent with the position taken here [on publicity] .... The book is not likely to be reviewed on "Sixty Minutes" or excerpted for the *Readers Digest.* My guess is that those who choose to read a book with this title are beyond saving anyway. More seriously, there are subjects that those interested can read, think, and talk about which are best not made part of candid, wide-ranging popular debate or explicit legislative decisions ([28], pp. 310—311).

In these arguments, Rhoads echoes the concern expressed earlier by Guido Calabresi and Philip Bobbitt: the cost of publicity in the realm of "tragic choices" is likely to be very high [5]. Without arguing for dishonesty, Calabresi and Bobbitt recognize that the usefulness and social acceptability of many allocation schemes depend on "the charade that they serve the purposes that they say they do" ([5], p. 24).

This "charade", according to the authors, serves many purposes. For example, obscurity about the true bases for allocating scarce lifesaving resources serves to mollify what might otherwise become strident conflicts over moral values. Thus, if elderly persons are disfavored in rationing because their more costly care is likely to result in fewer productive years of life, then it is less destructive of society's generally affirmed values to "hide" the true nature of the selection process. It might be claimed, for example, that such decisions are based solely on medical criteria. The trouble with publicity is that some of society's dearest values such as the belief in the pricelessness of each human life may be placed in jeopardy.

A further problem with publicity, according to Calabresi and Bobbitt, is that public knowledge of the tragedy of microallocation decisions is likely to focus criticism on the macroallocation decisions that have neglected to alleviate the scarcity of the resource in question.[6] Using the well-known example of maintenance hemodialysis for end-stage renal disease, these authors point out how unhappiness with (and lack of confidence in) microallocation decisions led to review of the macroallocation decisions ([5], pp. 181–191). Although different societies have approached this problem in different ways, it is not uncommon for awareness of the tragic microallocation decisions to generate pressure to make more treatment slots in artificial kidney programs available. But moving the decision to the macroallocation level does not avoid the tragedy of the choices. Even as one group, in this case those with kidney disease, gains the advantage of more available care, some other group becomes less likely to gain such favor. And depending on factors such as the emotional impact of the micro-allocation decisions, public knowledge of the details of such rationing may foster popular pressure to make very inefficient macroallocation decisions. As such inefficiencies come to light, the macroallocation decisions are subjected to further criticisms. The programs that *are* funded are compared for utility with those that *might have been*. The more society gains of this kind of knowledge of Good and Evil trade-offs at all levels of the rationing process, the more society may feel compelled to make highly inefficient decisions *if* it is to maintain its traditional commitment to the inestimable value of human life.

The problems just discussed, (1) the demoralizing effect on society that could result from weakening of the commitment to save every life possible, and (2) the pressure to make unwise macroallocation decisions, appear to head the list of objections to publicity for the rationing of medical care. But one other deserves at least brief mention.

The criticism could be made that the whole notion of publicity assumes something that is far from the truth: society (or some group within it) *owns* some scarce resources that are now to be distributed in some publicly acknowledged way. But just who owns, say, scarce artificial organs or the skills for transplanting such organs? Rather than supposing that society owns such resources, would it not be better to say that they belong to the persons who are in rightful possession of them and that they should be distributed, just as most other goods are, through fair exchanges in the market? Would not the market eliminate the need for publicly announced rules of allocation?

IV. ANSWERING OBJECTIONS TO PUBLICITY

How should those, who, like myself, favor publicity for rationing medical care respond? Without repeating all of the reasons for publicity given in Section II, I would like to sketch some possible answers.

First, it must be admitted that the predictions of publicity's effects are difficult to assess. (How much publicity have we actually tried?) I can see little point, however, in attempting to claim that publicity for rationing would not be costly. What must be said, as I see it, is that publicity *may* be expensive, but it is also necessary in a democratic society. If we are primarily concerned about the costly consequences of publicity, then we should also attend to the likely costs of non-publicity. Indeed, some of the best arguments for publicity (just as is true for democracy) stem from consideration of the negative aspects of the alternatives.

For example, Calabresi and Bobbitt acknowledge (more than Rhoads appears to) that the effectiveness of deceptive approaches to tragic decisions depends on the assumption of honesty and is, thus, fundamentally unstable ([5], pp. 24—25). The subterfuge is always just waiting to be unmasked. And this unmasking is not likely to be the work of only a few enterprising reporters or politicians, as Rhoads seems to suggest. Rather, given the gravity of the issue, rationing medical care is also likely to win the attention of scholars and others who are curious about the way society actually works. (Not to mention the more than idle interest of the potential recipients and their families.)

In the 1970s, for example, studies were conducted on the process of patient selection for dialysis [14]. The results were not always comforting. More recently, John Kilner has studied again the way in which people with kidney failure are either accorded or denied access to treatment [16]. Kilner's survey netted responses from the medical directors of 373 dialysis facilities and 80 kidney transplantation programs. These directors reported that a number of factors may influence a patient's chances of receiving either a transplanted kidney or dialysis. Contrary to the widely-held belief that everyone who needs treatment for end-stage renal disease is given such care under the provisions of the 1972 federal law, Kilner has found that many people may be refused treatment for a variety of reasons. For example, in addition to criteria such as the probability of medical success, a significant number of the medical directors reported that such factors as psychological stability, social worth, and ability to pay may have some weight in the treatment decisions.

When such studies are done, the rules, either implicit or explicit, that govern rationing become more apparent and, thus, more subject to criticism. (Or, if there are no observable rules, the rationing process may be open to the criticism of being capricious.) Perhaps we can imagine a society in which studies uncovering the actual bases for rationing are not permitted. But few of us, I should think, would choose to live in such a society. We would rather bear the costs of publicity.

And there is no denying that among those costs are likely to be the expensive results of pressure on society's policy makers to alter macro-allocations. For example, when journalists like Shana Alexander [23] publicized the nature of the rationing decisions for maintenance hemodialysis, the result was an outburst of public opinion that eventually led to a very expensive, national treatment program. It may be, of course, that the experience with End-Stage Renal Disease was atypical, and that other diseases with other types of high-cost treatments would not call forth the same kind of response. And, it might be added, not all other developed countries have gone the same route with dialysis.

Henry Aaron and William Schwartz, in their widely-read work on rationing medical care, for example, compare the British and American practices [1]. Americans, it turns out, receive dialysis and kidney transplantation at greater than three times the rate of their British counterparts. But, at least until recently, there has not been widespread public discontent in Great Britain concerning the rationing of treatment for kidney disease. Why?

The answer is, no doubt, complicated. But at least part of the explanation must be attributed to *non*-publicity. Aaron and Schwartz report a multitude of ways in which the British limitations on treatment for kidney failure are hidden from public view. For example, even though it has been rare for Britons over the age of fifty-five to receive maintenance hemodialysis, age is not explicitly or officially designated as a reason for exclusion. Indeed, according to Aaron and Schwartz: "Remarkably, few of the criteria for rejection are explicitly stated" ([1], p. 37). Nor does exclusion from dialysis generally represent any conscious scheme on the part of British physicians. Rather, once the budgetary limits have set the cap on available resources, physicians apparently tend to adjust their assessments of the criteria of medical suitability. And British patients and their families tend to trust their physicians' judgments even if the physicians conclude that aggressive treatment of the illness is not indicated. All this might strike many Americans as quite remarkable. Perhaps even more remarkable, however, is the fact that up to the time that Aaron and Schwartz concluded their research very few articles in the British press questioned this approach to rationing.

But all this may be changing. Within the past year, the differences between British rates of treatment for End-Stage Renal Disease and the rates of other countries have been much more widely publicized in Britain. (One wonders about the influence, in this process, of an impressive study like the one by Aaron and Schwartz.) An essay in the leading British medical journal, *Lancet*, claimed that hundreds of people with kidney disease die unnecessarily each year for want of adequate treatment [7]. And, as might be expected, pressure is now building to increase the budget for dialysis.

So it would seem that publicizing approaches to rationing often is likely to induce changes in macroallocation. And, inevitably, *some* of these changes will be miserably inefficient; the programs with better press, in some cases, will take funds that might have been spent elsewhere with greater lifesaving or health-enhancing effects.

But how much should this likelihood count against publicity? It is certain that not all such publicity-induced changes in macroallocation will be inefficient. Aaron and Schwartz themselves acknowledge, for example, that large amounts of money are allocated for some British medical treatments, such as chemotherapy for people with metastatic cancer, that are far less costworthy than *some* increase in expenditures for dialysis would be ([1], p. 89). One of the obvious, potential benefits of publicity lies in its capacity to reduce the probability that society will settle too easily, and with too little debate, for a macroallocation scheme that makes rationing necessary when it should not be. And even if the pressures generated by publicity do lead to some gross inefficiencies, is this not simply one of the hazards common to an open, democratic society? Moreover, is not publicity of the inefficiencies likely to be essential to their eventual correction? (Consider, for example, increased public awareness concerning needless surgeries.)

Granted, the debates about rationing, engendered by greater publicity, *may* be demoralizing. But are such disputes really more demoralizing than conflicts over other social issues, say, capital punishment or the appropriate level of expenditures for nuclear weapons? (At least with debates about rationing medical care, the focus is on the tragic limits we face in our efforts to *save* lives.) Many social issues have the potential for fostering demoralizing conflicts. And it is generally just these issues which are most in need of full public debate.

Up to this point, I have assumed that at least some scarce medical resources will be considered the products of social cooperation. I have also assumed that our society will continue to provide at least some types of scarce life-saving care to people in need when they are unable to pay for such care. For these reasons, the rationing of such resources presumably would be the legitimate subject of social debate and control. These assumptions comport well with current practice in our society, as I understand it. But what if the assumptions were wrong? For example, what if all such rationing were left entirely to the free market? Would such a market in scarce medical care not obviate the need to publicize rules for rationing?

The answer, it seems to me, is "no". As I am using "rules" (that is, the norms of any specifiable system of rationing), the formal principle requiring publicity should apply with as much force to the market approach as to any other. So, for instance, the rule might be that scarce life-saving medical resources are always made available to the highest bidder. Under this rule, when a few clinically successful, totally implantable artificial hearts become available, prospective recipients might be asked to submit sealed bids for the expensive devices and for the service of implanting and maintaining them. But, in all fairness, if this is the way that artificial hearts are to be made available, should not the rules be made known?

Perhaps such an example is too far-fetched. A more realistic illustration is simply the way consumers of medical care buy insurance. I take it to be a fundamental feature of our sense of fairness that the buyer should be made aware of the provisions of the insurance policy. Does the policy cover, for example, pediatric liver transplants? Or, if a person is purchasing care from a health maintenance organization (HMO), how will that organization handle rationing of kidney or heart transplantation? Surely, the purveyors of such care should be obligated to tell prospective customers what the rules for rationing are. Without such information, consumers could hardly make rational decisions about which competing plan to buy. And, indeed, one of the standard criticisms of the market in medical care is that relevant information is so often inaccessible to the consumer. So, advocates of a competitive model for medical care should be pleased with the federal government's recent decision to publish statistical information about the relative quality of care in hospitals throughout the nation [22]. And it should be hoped that this sort of commitment to publicity will be extended to rationing, especially if we are to rely more heavily on the market approach for future rationing of medical care.

I certainly am not calling for such a market approach to rationing. For reasons I have stated elsewhere, I do not favor complete reliance on the market in this area ([34], pp. 155-158; for similar arguments see [5], pp. 83—127 and [15], pp. 179—188). What I *do* favor, and what I believe is far more likely to be socially acceptable because of its greater compatibility with our society's commitment to a certain basic equality, is an approach similar to that recently proposed by the Massachusetts Task Force on Organ Transplantation. The Task Force's report recommends screening candidates first to determine the probability of medical success, interpreted to mean the likelihood of "living for a significant period of time with a reasonable prospect for rehabilitation" ([26], p. 11). If, after the appliciation of medical criteria, there are still more candidates than can be served by the limited resources, then, according to the Task Force's recommendation, those for whom the treatment is likely to be beneficial should be selected on a *first-come, first-served* basis. The report explicitly rejects such selection criteria as estimates of the candidates' social worth or ability to pay.

I mention these points now merely for background. From the

standpoint of the present essay, what is particularly noteworthy about the Massachusetts report is its clear call for publicity. In the words of the report: "Patient selection criteria should be public, fair, and equitable. Primary screening should be based on medical suitability criteria made available to the public . . ." ([26], p. 11). And in order to help insure that the medical criteria are equitable and do not hide what may be unfair bases for patient selection, the report suggests that the criteria "be reviewed and approved by an ethics committee with significant public representation, filed with a *public* agency, and made readily available to the public for comment" ([26], p. 20).

Implementation of the Massachusetts report, if such occurs, will represent a genuine departure from previous practice. Increased publicity will not insure that justice is achieved in Massachusetts, or anywhere else. But, if my argument is correct, the report's prescription for publicity deserves praise.

V. CONCLUSION

Many questions remain. Worrisome, for example, is the potentially corrosive effect of publicity on the morale of medical practitioners. Little has been said here about the changes that rationing might cause in medicine's traditional values of personal, individualized care, changes which may bring about pain that publicity may only serve to intensify. But I am inclined to think that any demoralizing effect of publicity depends not so much on the practice of publicizing the rules as it does on the types of rules that are publicized. (One reason to think that the Massachusetts Task Force's proposal for publicity may succeed is that the rest of their recommended approach to rationing probably will *bear* being publicized.) In the end, we cannot eliminate many of the distressing costs of rationing medical care. But publicity should help us bear these burdens together. And publicity may finally be humanizing in one more way: We will be brought face to face with that most human of all traits, knowledge of our own finitude.[7]

Walla Walla College
College Place, Washington

NOTES

[1] Readers familiar with the work of Guido Calabresi and Philip Bobbitt will notice my dependence, in much of this essay, on their impressive work [5].

[2] It is not my intention here to enter the debate about what constitutes a "decent minimum". But it does seem quite clear that the belief that what is required of society is the provision of such a minimum (and not the best care or equal care for all) currently has achieved the status of hegemony. If one of the purposes of ethics is to bring coherence and structure to the already developing beliefs and practices of a society, then, it seems to me, high marks for prescience should go to Charles Fried [10]. By the mid-1970s, Fried was calling for an honest admission that it was foolhardy to offer a universal right to the best medical care. What we should do, according to Fried, is settle for the provision of a "decent minimum", or respectable floor of medical care beneath which no citizen should have to fall.

[3] In past work, I have used the term "triage" to refer to the rationing of scarce life-saving medical resources [34]. But I have become convinced that "triage," owing to its origins in military and disaster medicine, may come attached to some unhelpful associations. James Childress has argued, for example, that "triage" should not be used in place of expressions such as "rationing" or "allocation" [6]. Childress is particularly concerned that the metaphorical usage of "triage" in contexts, such as neonatal intensive care, far-removed from the term's battlefield origins may predispose decision-makers to follow utilitarian principles that could be morally unacceptable. In response to this concern, one could point out that "triage" has been used historically to refer to processes of patient selection in many areas of medicine. And, it might also be said, "triage" continues to be used for a wide variety of approaches to rationing medical care, and not just utilitarian ones. (See, for example, [18]). However, since I have no particular commitment to the usage of "triage," and since I certainly do not want to use a term that might bring to mind only utilitarian approaches to rationing, I have decided to abandon the term in this essay.

[4] I say *middle* school because, if Jean Piaget's work is reliable, children at about that age typically go through an interesting transition in their attitudes toward rules [23]. The very young child imagines that rules, even those for children's games such as marbles, were laid down by some authority — maybe parents, grandparents, or God. But, in spite of the young child's belief that some absolute authority made the rules, they are often misunderstood, misapplied, or simply ignored. As children become older, they are more likely to think of rules as the result of mutual consent. (So, for example, older children sometimes will enter into elaborate negotiations of the rules, a process that may take longer than the game itself.) When the children begin to think of the rules as resulting from their own agreements, the rules are much more likely to be understood and followed. In Piaget's terms, this transition is from a heteronomous morality of *constraint* to an autonomous morality of *cooperation*.

[5] In addition to Plato's remarks ([24], pp. 93—95; 414a—415d), I am thinking of Machiavelli's discourses on religion: "The rulers of a republic or a kingdom . . . should uphold the basic principles of the religion which they practice . . . and, if this be done, it will be easy for them to keep their commonwealth religious, and in consequence, good and united. They should foster and encourage everything likely to be of help to this

end, *even though they be convinced that it is quite fallacious*" [emphasis supplied] ([24], pp. 142—144).

[6] Calabresi and Bobbitt use the expression "first-order decision" for macroallocation and "second-order decision" for microallocation.

[7] I would like to thank my colleagues, Hollibert Phillips and Charles Scriven for useful comments on an earlier draft of this paper, and I would like to thank Lee Johnston, reference librarian, for valuable help in gathering material for the paper.

BIBLIOGRAPHY

[1] Aaron, H. and Schwartz, W.: 1984, *The Painful Prescription: Rationing Hospital Care*, The Brookings Institution, Washington, D.C.

[2] Alexander, S.: 1962, 'They Decide Who Lives, who Dies: Medical Miracle Puts a Moral Burden on a Small Committee', *Life 9*, 102ff.

[3] Baier, K.: 1958, *The Moral Point of View: A Rational Basis of Ethics*, Cornell University Press, Ithaca.

[4] Cahn, E.: 1949, *The Sense of Injustice*, Indiana University Press, Bloomington.

[5] Calabresi, G. and Bobbitt, P.: 1978, *Tragic Choices: The Conflicts Society Confronts in the Allocation of Tragically Scarce Resources*, W. W. Norton and Company, New York.

[6] Childress, J.: 1983, 'Triage in Neonatal Intensive Care: The Limitations of Metaphor', *Virginia Law Review 69*, 547—561.

[7] Dietch, R.: 1984, 'Commentary: UK's Poor Record in Treatment of Renal Failure', *Lancet 2*, 53.

[8] 'Due Process in the Allocation of Scarce Lifesaving Medical Resources', 1975, *The Yale Law Review 84*, 1734—1749.

[9] Fox, R. and Swazey, J.: 1974, *The Courage to Fail: A Social View of Organ Transplants and Dialysis*, University of Chicago Press, Chicago.

[10] Fried, C.: 1976, 'Equality and Rights in Medical Care', *The Hastings Center Report 6*, 29—34.

[11] Fuchs, V.: 1984, 'The "Rationing" of Medical Care', *The New England Journal of Medicine 311*, 1572—1573.

[12] Gunby, P.: 1983, 'Media-Abetted Liver Transplants Raise Questions of "Equity and Decency"', *JAMA 249*, 1973—1982.

[13] Kant, I.: 1983, *Perpetual Peace and Other Essays*, T. Humphrey, tr., Hackett Publishing Company, Indianapolis.

[14] Katz, A., and Proctor, D.: 1969, *Social-Psychological Characteristics of Patients Receiving Hemodialysis Treatment for Chronic Renal Failure* (mimeographed), Public Health Service, Kidney Disease Program, Rockville, Maryland.

[15] Katz, J., and Capron, A.: 1975, *Catastrophic Diseases: Who Decides What?* Russell Sage Foundation, New York.

[16] Kilner, J.: 'Who Receives Scarce Resources? An Empirical and Ethical Study', *The Annual of the Society of Christian Ethics*, in press.

[17] Machiavelli, N.: 1975, *The Discourses*, L. J. Walker, tr., Routledge & Kegan Paul, Boston.

[18] Margolis, J.: 1985, 'Triage and Critical Care', in L. Kopelman (ed.), *Ethics and Critical Care Medicine*, D. Reidel Publishing Company, Dordrecht, Holland.
[19] Mechanic, D.: 1978, 'Rationing Medical Care', *The Center Magazine 11*, 22–31.
[20] Menzel, P.: 1983, *Medical Cost, Moral Choices: A Philosophy of Health Care Economics in America*, Yale University Press, New Haven, Connecticut.
[21] 'Patient Selection for Artificial and Transplanted Organs', 1969, *Harvard Law Review 82*, 1322–1342.
[22] Pear, R.: April 16, 1985, 'U.S. to Offer Consumers Access to Data on Health Care Quality', *The New York Times*, p. 9.
[23] Piaget, J.: 1965, *The Moral Judgment of the Child*, M. Gabain (tr.), The Free Press, New York.
[24] Plato: 1968, *The Republic*, A. Bloom, tr., Basic Books, Inc., New York.
[25] Rawls, J.: 1971, *A Theory of Justice*, Harvard University Press, Cambridge.
[26] 'Report of the Massachusetts Task Force on Organ Transplantation': 1985, *Law, Medicine and Health Care 13*, 8–26.
[27] Retting, R.: 1976, *Health Care Technology: Lessons Learned from the End-Stage Renal Disease Experience*, Rand Corporation, Santa Monica, California.
[28] Rhoads, S.: 1980, 'How Much Should We Spend to Save a Life?' in S. Rhoads (ed.), *Valuing Life: Public Policy Dilemmas*, Westview Press, Boulder, Colorado.
[29] Schwartz, H.: 1983, 'Access, Equity, and Equality in American Medical Care', in President's Commission for the Study of Ethical Problems in Medicine and Bomedical and Behavioral Research, *Securing Access to Health Care, Volume Two: Appendices: Sociocultural and Philosophical Studies*, United States Government Printing Office, Washington, D.C., pp. 67–78.
[30] Sidgwick, H.: 1966, *The Methods of Ethics*, Dover Publications, New York.
[31] Starr, P.: 1983, 'Medical Care and the Pursuit of Equality in America', in President's Commission for the Study of Ethical Problems in Medicine and Biomedical and Behavioral Research, *Securing Access to Health Care, Volume Two: Appendices: Sociocultural and Philosophical Studies*, United States Government Printing Office, Washington, D.C., pp. 3–22.
[32] Wallis, C.: 1982, 'Which Life Should be Saved?', *Time 120*, 100–101.
[33] Wildavsky, A.: 1977, 'Doing Better and Feeling Worse: The Political Pathology of Health Policy', in J. Knowles (ed.), *Doing Better and Feeling Worse: Health in the United States*, W. W. Norton Company, New York, pp. 105–123.
[34] Winslow, G.: 1982, *Triage and Justice: The Ethics of Rationing Life-Saving Medical Resources*, University of California Press, Berkeley.

DAVID D. FRIEDMAN

COMMENTS ON "RATIONING AND PUBLICITY"

The essential point of Gerald Winslow's paper is that, while there may be serious costs associated with making public the rules by which medical care is rationed, those costs are worth bearing. While the author asserts that his arguments apply to both market and governmental systems for allocating medical care, most of his discussion is in terms of political allocation in a democratic society.

In the first part of this commentary, I will try to show that neither the arguments nor the conclusion apply to a market system — that Winslow, viewing medical care from the perspective of a political system, has described a set of problems which result from trying to provide medical care politically. In the second part of the commentary, I will approach the question from a diametrically opposite direction. Where Winslow looks at market provision as a special case of political provision, I will try to view political provision as a special case of market provision. I will argue that, looked at in that way, the "rules for rationing life-saving medical care" are not facts, which we may choose to admit or conceal, but myths, which we may or may not wish to create.

I. MARKET HEALTH CARE AND PUBLICITY

At the end of part III of his paper, Winslow asks, "Would not the market eliminate the need for publicly announced rules of allocation?" His answer is "no". While he does not support a purely market system, he claims that, for those who do, the arguments for publicity would still apply. He does not try to redo all of his arguments in the context of the market, but he does give two examples. One is the case in which the rule for allocating a scarce life saving resource — an artificial heart — is that it goes to the highest bidder; he asserts that if this is the rule, then, "in all fairness", it should be made known. The other example is that of an insurance policy or an HMO; surely, he argues, "the purveyors of such care should be obligated to tell prospective customers what the

G. J. Agich and C. E. Begley (eds.), The Price of Health, 217–224.

rules for rationing are", in order that the customers will know what they are getting for their money.

In discussing this issue, it is important to distinguish between two very different sorts of rules. On the one hand, we have the fundamental rules determining who is entitled to decide what — the rules that define and allocate property. These rules define who owns what, what an owner can do with what he owns (for example, ownership of a revolver does not imply the right to shoot it in the direction of other people), how ownership is transferred, and so forth. While the details of who owns what are occasionally kept secret in a market system, the overall framework of rules is, and presumably must be, public, in order that all of the participants may know what they can or cannot legally do. In this sense, the requirement of publicity applies to the market as well as to the political system.

But it seems clear, from Winslow's examples, that these are not the sorts of rules he is thinking of. "Allocate an artificial heart to the highest bidder" is not a rule of the market — it is a policy that a particular market participant (the owner of the artificial heart) happens to be following. Within the same set of market rules, that owner might follow any of a great variety of policies. If the owner were an individual, he might allocate the heart to the individual he thought most worthy, to someone he liked, to the highest bidder, at random, or in whatever other way he chose. If the owner were a for-profit hospital, it might choose to reject the highest bidder in order to get good publicity by giving the heart to a particularly attractive donor — say an appealing and photogenic child.

In such a market system, the "rules" that determine how medical care is allocated are simply the preferences of the numerous individuals (and firms) that control the relevant resources. It is hard to see why any of the six reasons which Winslow cites in favor of publicity would imply an obligation on the part of all of those individuals to state publicly the grounds on which they were acting.

Even in the case of a health insurance company or HMO, while it seems likely that such an organization would choose to state its policies in order to persuade customers to buy its products, one can imagine legitimate reasons why it might not. Suppose that such a firm has worked out a policy which, in its opinion, provides its customers better value for their money than the policies used by its competitors — perhaps by spending more on prevention and less on cure, economizing in non-obvious ways, or whatever. The firm might well regard the details of its policy as a valuable trade secret. Its refusal to divulge those details would make it more difficult for it to attract patients, but that difficulty might be overcome in other ways — by good word-of-mouth reputation, by guaranteeing output (for example, the health insurance might also include life insurance, thus imposing a sizable cost on the firm if the patient died), or in some other way. As long as the firm does not mislead potential customers as to what it is offering, it is hard to see any reason of principle for requiring it to make all of its policies public.

In exploring the application of his argument for publicity to a market system of health care, Winslow is, I think, handicapped by the fact that he does not really believe the market to be an appropriate way of allocating health care. One way of avoiding this problem is to consider a different market, one to which his arguments, stated in the abstract, seem to apply just as well as they do to the medical market, but in which virtually everyone regards market allocation as appropriate. If, in that context, the conclusion — that the rules which participants use to allocate the resources they control ought to be public — seems inappropriate, we may take that as evidence that the same conclusion would be inappropriate if medical care were allocated entirely on the market.

The Other Market for Hearts

The market I will consider is the marriage market — the system by which, in our society, husbands are allocated to wives and wives to husbands. The services being allocated, like those allocated in the medical market, are of enormous value to the individuals concerned. Just as in that market, the decisions of one participant may impose large costs on another. The one crucial difference is that, while very few people believe that the allocation of medical care should be left entirely to the free market, most people believe that the allocation of spouses should be.[1]

The rules defining property rights on the marriage market are simple and public. Each individual belongs to himself, and a marriage occurs if and only if the two individuals concerned consent to it.[2]

In discussing the medical market, Winslow asserts the claims of publicity not merely for the rules defining property rights but also for the rules that determine the actual outcomes — what I earlier described as the policies of the participants. The equivalent of that on the marriage market would be to require every individual to post a set of desiderata for a potential spouse — and abide by it. The rejected suitor could presumably demand an explanation of why he was rejected, and sue — in the modern, not the old, sense of the term — for the lady's hand if he found the explanation unsatisfactory.

I suspect that most readers of this essay will agree that such a set of requirements would be absurd. They take it as given that individuals own themselves, and hence that while the refusal of the lady I love to marry me may cause me great distress, that fact gives me no claim to her. Both the decision of whether to marry me and the decision of whether to explain her reasons are entirely up to her.

The requirement of publicity on the marriage market seems inappropriate because we believe that a bride owns her body. To those who believe, similarly, that nurses, surgeons, and all of the other participants in the medical market own their own bodies — and what those bodies create, and what is transferred to them by other legitimate owners — the requirement of publicity on the medical market must seem equally inappropriate.

II. IS NATIONAL HEALTH POLICY A USEFUL MYTH?

So far, my purpose has been to show that while Winslow's conclusion may be correct for a political system for allocating health care, his arguments do not apply to a market system. In the remainder of my comments, I will try to use my discussion of a market system as a base from which to examine political systems. My objective is to criticize not his conclusion but his question.

Winslow takes the existence of rules for allocating medical care for granted; in his words: "Presumably, if rationing is becoming more obvious, then some people should know how it is being conducted." The first question I wish to raise is whether, or in what sense, such rules exist.

One could claim that every system has rules; some outcome occurs, and presumably something determines what that outcome is. In this sense, even the weather has rules; there is some reason why it rained today, whether or not we know it. But when Winslow talks about rules for allocating medical care, it is clear that he means explicit rules which some people know and could explain to others. It is by no means clear, even in a political system, even an entirely political system such as the British National Health Service, that rules in that sense exist.

In such a system there are, typically, rules and regulations written down somewhere, but that does not mean that knowing those rules would allow you to predict what the system will do. As Mary Ann Baily points out: “Even in the ‘socialized’ National Health Service, there is no single authority making all the decisions. Instead, a system of interlocking levels operates with different mechanisms at different levels.” In such a system, the treatment that a patient actually gets will depend on many things other than the formal rules of the organization.

For example, a patient who happens to be a famous public figure is likely to get more attention than one who is not, if only because hospital authorities will want to avoid the risk of bad publicity. A patient who is the child or parent of a physician on the staff of a hospital may well get better treatment than one who is not. A cute and affectionate child is likely to get more attention and better treatment than a grumpy old man.

The question is not whether these things should happen but whether they do. The essential fact is that the pure centralized bureaucracy, in which all of the orders come from the top and everyone below merely carries them out, is a work of fiction — and an implausible one at that. Nobody at the top of a large organization knows enough to make all of the decisions, hence all such organizations are, at least to some degree, decentralized. It follows that the actual outcome produced by a relatively centralized system such as the NHS, like the outcome produced by a market, depends on the decisions and preferences of a large number of different individuals, many of whom will not, and perhaps can not, state explicitly the grounds for all of their choices. Hence it follows that rules in the sense in which Winslow discusses them — explicit rules determining how health care actually gets allocated — do not exist.

Having recognized that, we may then restate Winslow's question in any of three different ways. We might ask whether the explicit rules which apply to the allocation of health care, whether or not they determine what actually happens, should be made public. Alternatively, we might ask whether it is desirable to try to figure out how the system really works — to try to reconstruct the implicit rules, and thus develop a theory which predicts who will get what care — and, if we succeed, make the results public. As a final and interesting alternative, we might ask whether, even if rules do not exist, it is desirable to pretend that they do.

Winslow's discussion seems most relevant to the first form of the question. If explicit rules exist, it is at least possible to make them public. Since rules presumably affect, even if they do not determine, the outcome of the system, those who care about its outcome may be interested in knowing, and perhaps trying to change, those rules. Hence, we may interpret Winslow's arguments for publicity as arguments for making public the formal rules of the organizations involved in producing and allocating health care.

Parts of his discussion may also be taken as arguments for discovering and making public the implicit rules determining the allocation of health care. If one's model for the society is, as his sometimes appears to be, civics class democracy, in which the system is ultimately controlled by the informed decisions of the voters, then giving voters information about what they are actually getting in exchange for tax money spent on health care seems like a good way of producing better decisions.[3]

But not all of the arguments which Winslow makes in favor of publicity remain relevant in this case. The six arguments he lists all apply to rules, not to outcomes. Some, such as argument 2, can be converted into arguments about outcomes by adding the observation that individuals approve of rules, in part, because of their opinion concerning the outcomes that those rules will produce. Others cannot.

Consider, for example, argument 5: "Knowledge of the rules not only establishes legitimate expectations, it also provides a basis for complaints if those expectations are unfairly abrogated". If the rules are simply a description of how, on average, the system works — a model to predict outcomes — it is hard to see how the fact that an outcome does not fit the rules could provide anyone with a basis for complaint. If the prediction turns out to be wrong, that is evidence that the model is not a perfect one, but it implies nothing at all about whether the outcome was fair, since the "rules" are merely an (approximate) description of what is, not of what ought to be.

Finally, we come to the most interesting of the three formulations of Winslow's question. Given that rules do not exist, is it desirable to pretend that they do? Is national health policy a useful myth?

I do not intend to answer that question here. I will, however, point out that some of Winslow's points, both for and against publicity, are relevant to that question as well. If the belief that the members of a society understand and approve of the rules under which it runs helps to make that society more stable and more attractive, it may do so even if the belief is false. If the realization that our society chooses to let some people die is destructive of the social fabric, it remains destructive even if the description of a society as "choosing" is a false and misleading metaphor, and the realization hence false.

One might even argue that, for purposes of publicity, a myth has advantages over a fact. In order to publicize the *fact* that everyone in our society receives adequate medical care, one must first make that fact true, which is expensive and perhaps impossible. To publicize the *myth* is easier and less expensive — and, as Mary Ann Baily points out in one of the notes to her paper, has already been done.

Some will reply that facts, being true, are harder to debunk, hence longer-lived, than myths, and that making them public therefore yields a larger return. Whether this is true is an empirical question. An energetic investigator could probably compile, from the history of recent decades, an extensive list of facts that have been successfully debunked and myths that have not.

A.B. Freeman School of Business
Tulane University
New Orleans, Louisiana

NOTES

[1] There are some exceptions to this principle, such as incestuous or homosexual marriages.

[2] This is a feature of modern Western society; in many other times and places it was the consent of the parents that was the essential condition, at least for the bride. One may suspect that while the consent of the bride was not a sufficient condition in traditional societies, it was generally a necessary one if the marriage was to be happy. As an extreme example, consider the fate of Hallgardr's first husband in *Njal Saga*. She was married to him against her will and resolved the problem by arranging to have him killed.

[3] For a detailed explanation of why I do not consider civics class democracy a very useful model for understanding any large society, see [1], Chapters 17 and 18. The essential problem is the public good problem as applied to voters. A voter who bears any significant cost in order to discover and support "good laws" is producing a public good; he pays all of the cost and gets a tiny fraction of the benefit. The result is that the rational voter is typically ignorant, except when he is part of a concentrated and

organized interest group. Legislation is then the result, not of the rational deliberation of voters seeking the general good, but of a political market in which groups compete to "buy" legislation that benefits their members, usually at the cost of non-members.

BIBLIOGRAPHY

[1] Friedman, D.: 1986, *Price Theory: An Intermediate Text*, South-Western Publishing, Cincinnati.

PART IV

CONTROLLING COSTS/MAXIMIZING PROFIT: THE ROLE OF PROVIDERS

CHARLES E. BEGLEY

PHYSICIANS AND COST CONTROL

I. INTRODUCTION

It can be argued that there are only two policy alternatives for the United States regarding the integration of cost control into the health care system: first, rationing by government, private payers, and health care plans through administrative rules, or second, rationing by providers and patients in response to economic incentives. In the first approach the rules of resource allocation are defined outside the provider-patient relationship on the basis of institution- or society-determined resource constraints. Certain service options are not made available, and providers and patients are made to choose what other services will be provided on the basis of the allocated resources. This approach is exemplified by the regulatory policies of the 1970s, which attempted to contain costs by imposing limits on the availability of resources and establishing guidelines for resource use.[1]

In recent years policies which emphasize the incentives approach to cost control have become more popular. This strategy relies on economic rewards or sanctions for care at less cost.[2] The federal government is leading the way with the introduction of DRGs (Diagnosis Related Groups), which, by paying hospitals a fixed sum per patient, creates incentives for hospital management to press physicians to order fewer tests and procedures and for early discharge of patients.

States and businesses are following suit through efforts such as contracting with PPOs (Preferred Provider Organizations) for the care of patients. PPOs are groups of health care providers who compete for contracts with payers by offering discounted fees. Policies to encourage enrollment in alternative delivery systems such as HMOs (Health Maintenance Organizations) and primary care case-management systems are also becoming more common. The cost control potential of HMOs rests on the idea that providers who have an economic stake in these organizations have an incentive to hold costs down in order to meet the medical care needs of a defined population of patients for a prepaid fixed price. Also, cost control incentives are created through increased

G. J. Agich and C. E. Begley (eds.), The Price of Health, 227–244.

cost-sharing by patients. This strategy consists of changes in health insurance benefits which increase the expense borne by the patient at the time of treatment. By putting the patient at financial risk, this strategy indirectly creates incentives for providers to reduce costs.[3]

While applauded for allowing greater autonomy within the provider-patient relationship, the use of economic incentives is often criticized for creating ethical conflict for medical decision makers. In this paper I consider the arguments against this approach from the point of view of physicians and other providers. Specifically, I examine three issues which make assessing the ethical implications of these policies difficult: first, because of uncertainties regarding the benefits of medical care, it is unknown whether cost considerations pose a serious threat to the quality of care. In other words, it is not known precisely when or how often cost concerns will conflict with concerns about patient well-being. Second, if conflict does arise, it is not clear what values should be followed. The traditional view of medical ethics suggests the primacy of the individual patient, yet utilitarian theory and distributive justice considerations suggest that the individual patient's choice of care may be overridden by societal concerns in some instances. Third, conflicts regarding informed consent complicate this debate because they illustrate the potential danger which rationing poses for the fiduciary character of the traditional patient-physician relationship.

Before undertaking this discussion, however, it is useful to note that the debate over policies which rely on economic incentives to achieve cost control is by no means limited to problems of provider ethics. The reimbursement schemes and organizational arrangements which have been developed, or are proposed, present a variety of ethical problems regarding the behavior of patients, insurers, employers, and health care institutions, as well as providers.[4] The appropriateness of the use of economic incentives versus other incentives will, of course, be based on different factors in each of these contexts. My discussion is focused not on the methodological or empirical issues raised in applying economic analyses to medical decisions, but on the question of the appropriateness of having physicians adopt economic criteria in deciding what medical services to offer their patients. I am more concerned with the assumptions underlying the ethical arguments against the use of these criteria in clinical decision making than the analytical techniques or methods of measurement regarding their application.

II. TWO VIEWS OF PHYSICIAN OBLIGATION

Many physicians are reluctant to accept responsibility for cost in clinical decisions. For example, in a recent commentary in the *New England Journal of Medicine,* Norman G. Levinsky stated: ". . . physicians are required to do everything that they believe may benefit each patient without regard to costs or other societal considerations. In caring for an individual patient, the doctor must act solely as that patient's advocate against the apparent interests of society as a whole" ([18], p. 1573). According to this view, the physician is obligated to provide all care that is beneficial to his patient based on the principle of beneficience. There is concern that, by putting physicians in a position of having to balance patients' desires and needs against the obligation to contain costs, they could not remain advocates and agents for their patients. Arnold S. Relman expressed that very concern in a recent comment on DRGs:

> . . . they raise a disturbing ethical question about the role of the medical profession. Physicians have an ethical and legal responsibility to do the best they can to ensure that their patients receive all necessary medical care. When medical needs are not clear, patients should receive the benefit of the doubt. Is it in the patient's interest for his physician to have an economic stake in reducing his hospital services? ([29], p. 109).

Many believe that medical ethics supports this viewpoint, arguing that the fiduciary relationship between doctor and patient demands that the patient's good be pursued regardless of cost.[5] Edmund Pellegrino has articulated a strong version of the physician's obligation to serve his patient: "The prime focus of the physician's intention is the good of that individual patient, not some distant patient, not the good of society, or the greatest good of the greatest number . . ." ([27], p. 9). Pellegrino maintains that the physician is always obliged to advocate his patient's welfare — the only exception being the case in which that welfare is immediately and urgently a danger to others, for example, the patient with a contagious disease or the dangerous psychotic ([27], p. 9).

Pellegrino's remarks are based on a particular view of the medical profession which has considerable support. According to this view, physicians are to use their medical knowledge, skills, and power to achieve a certain end, namely the elimination of disease in the service of the enhancement of patient autonomy. The problem of how best to spend a limited amount of resources is not an issue. All that has to be shown is that one treatment is better for the individual patient than another. The physician's obligation is to make that patient aware of the package of diagnostic or therapeutic services that best meets his needs. He has no obligation to those he does not treat. The third-party, fee-for-service payment system has reinforced this model allowing physicians and patients to pursue what Avedis Donabedian has called "absolute quality" — do all that is possible for the patient regardless of cost ([9], p. 278).

A problem with this view from an economic perspective is that it tends to encourage misallocation of scarce resources. With the development of more and more expensive techniques and devices that can slightly improve a prognosis or marginally prolong life, the expenditure that has to be made before reaching the "do all that is of benefit" stopping point has grown to almost unlimited levels. As medical costs continue to rise, questions are being raised about whether society can afford an ethic that allows providers to act as though the amount of money available for health care is unlimited.

Duncan Newhauser describes a case which suggests the kind of inefficiency that results from this ethic:

> One of the least cost effective interventions, I am told, is the treatment of long-term alcoholics with bleeding esophageal varices. This may require innumerable transfusions and two operations on the esophagus, plus a porta caval shunt. This allows the patient to die of kidney failure rather than by bleeding from the esophagus. At worst, it means a socially acceptable form of death. At best, it allows the alcoholic to return to the bottle for a few more months before death. The cost per case may approach $20,000 and the cost per year of life saved may approach a million dollars .... Once a junior house officer in the Emergency Department decides to accept such a patient, he precipitates the whole $20,000 expenditure. Ethically, local doctors do not feel they can let such patients bleed to death on the street; after admission they feel morally obliged to do all they can ... ([23], pp. 27—28).

There is concern regarding the distributional results of this ethic as well which has been high quality care for some and no care for others. Under the traditional view, the physician need not concern himself that as a result of the $20,000 expended in the above case, for example, someone may die of the lack of medical care, as long as that individual does not appear on the physician's doorstep.

The alternative view is that the physician should be required to incorporate efficiency concerns in medical decisions ([17], [38]). This view maintains that physicians have obligations not only to their patients, but also to society. It explicitly recognizes scarcity, pointing out that providing medical resources to manage one patient's problem may mean that another patient's needs will be ignored. It requires the physician to evaluate the costs and benefits of various treatment options in order to use resources efficiently. Under this viewpoint, the goal of the system is to maximize the health of a defined population given a limited amount of resources such as doctors' time, hospital beds, plasma, and pharmaceuticals. The focus of the concern is not limited to the care of a particular patient, but also considers the particular patient's needs in relationship to the needs of others and in light of available resources. As a result, this view may conflict with the traditional view of medical ethics since it may mean doing less than the maximum for some patients in order to conserve resources for the greater benefit of other patients.

In applying the efficiency view to clinical decisions, Milton C. Weinstein and William B. Stason suggest that the physician consider all direct and indirect (overhead) costs associated with the provision of a medical service, all induced costs associated with adverse side effects, and all savings in future costs due to prevention or alleviation of subsequent illness [40]. In this context cost is explicitly limited to its economic sense of scarce resources consumed.

On the benefit side, the physician must attempt to estimate the effects of medical care in commensurate units so that the benefits and costs of different uses of resources can be compared across patients and procedures. Without going into the complexities of their argument, Weinstein and Stason recommend Quality Adjusted Life Years (QALY) as a measure of benefit because it attempts to capture patient's preferences for life-extension as well as quality of life ([40], p. 717). This measure assumes that medical care should be evaluated in terms of its effect on health. In the context of their model, physicians should strive to maximize the expected number of life years resulting from a medical action, adjusted for improvement (or decline) in quality of life due to alleviation or prevention of morbidity.

The cost-benefit rule is illustrated by two general types of decisions that physicians typically face. The first concerns decisions to order diagnostic tests. The physician concerned with the efficient use of limited resources will base his test ordering decisions on answers to the following questions: First, will the test results lead to changed decisions? If no subsequent decision rests on the results of a test, then it has no value to the decision maker. Second, is the risk to the patient less than the expected gain (in life years adjusted for quality) resulting from the decision change? The difference between the expected survival rate associated with the optimal strategy excluding the test and including the test is the benefit of the clinical information. Costs are incorporated by asking if the cost of the test relative to the increased survival rate expected by conducting the test is acceptable. In answering this question, the physician must balance resource constraints against patient benefits. The provider ethic of ordering all tests which might possibly benefit the patient is replaced by an ethic of ordering only those tests whose benefits justify their costs.

The other type of decision involves evaluating the effects of various treatments and choosing the alternative which has the best expected outcome and costs. The decision rule is to perform all treatment options for which the analysis indicates an acceptable cost per year of quality adjusted life saved. Acceptability is determined on the basis of an existing budget constraint and/or the availability of alternative uses of resources that yield greater benefit per dollar cost. A procedure which may have some benefit for the patient may be rejected on the grounds that the benefits are not great enough to justify the costs. The provider ethic of maximizing quality of care to the patient irregardless of cost is replaced by the ethic of limiting care to that which is cost-beneficial from society's point of view.

Ethical conflict arises when the patient's perspective on the costs and benefits of a particular test or treatment procedure is different from society's. The classic example is the case of the sixth consecutive stool guaiac in screening for colon cancer. Duncan Newhauser and A. M. Lewicki estimated that the marginal cost of each new cancer found with the sixth test would be almost $50 million [24]. Placing any value whatsoever on preserving society's scarce resources, the physician will choose not to perform the test. But the patient may be sufficiently afraid of cancer that he may wish to pay $1 for the extra test. In cases such as this, the physician must balance conflicting values.

III. UNCERTAINTY AND COST CONTROL

In considering the traditional view based on beneficience and the efficiency approach to clinical decision making, it may be instructive to recognize that they conflict only when the medical services the doctor might order have clear and substantial benefit for the patient. If cost considerations lead to the elimination of services that have little or no benefit, for example, the laboratory test that has no influence on the management of the case or the unjustified elective surgical procedure, then the traditional view of medical ethics and the efficiency ethic would agree on the correct action. Indeed, to the extent that cost considerations reduce risks associated with unnecessary care, they may be a required part of ethical medicine. If cost considerations lead physicians not to perform tests or conduct procedures which have clear medical benefits, however, a conflict exists.

Some who are critical of policies which encourage providers to restrict medical services assume that most medical care has clear and substantial benefits for the patient and that efforts to reduce services will necessarily lead to lower quality of care. This view is widespread and illustrated in a recent article by E. Haavi Morreim [22]. Morreim argues that the current policy mix of DRGs, HMOs, etc., present major value conflicts for physicians. She offers as an example the question of how long a patient needs intensive care after an uncomplicated myocardial infarction. She suggests that the physician will be required to choose between society's concern for efficiency and the assurance that the patient will recover. But this assumes, as she does not seem to be aware, that intensive care for uncomplicated myocardial infarctions provides clear and substantial benefits to all patients. This assumption must be questioned, because it is not clear that intensive care benefits all the patients who receive it.

Another plausible explanation is that the extra days in the ICU provide no significant benefit, but, since the actual cost to the patient is minimal and the risks are minimal, the patient's or physician's convenience becomes a factor, economic incentives for the hospital and/or physician influence the decision, or some other non-medical variable becomes determinant. This is not to say that the service decisions made on the basis of these criteria represent "pure waste", or in Morreim's terms "sheer foolishness" ([22], p. 272). Rather, it seems that absent a budget constraint, variables besides medical indications are important determinants of medical practice. It is at least arguable that the introduction of economic incentives for cost reduction could reduce the importance of these factors without affecting the quality of medical care.

There are numerous empirical studies which suggest that much medical care is discretionary. The debate is over how much. Some of these studies show how widely treatment patterns for similar conditions vary from place to place, others show how much treatment is inconsistent with expert opinion on what is necessary, and a third group shows how non-medical factors influence patterns of treatment. Differences of up to threefold in the frequency of hospital admissions and average length of stay for comparable patients with comparable illnesses have been identified ([43], [44]). In a hospital study of England and Wales by M. A. Heasmen and V. Carstairs the median length of stay (LOS) for myocardial infarction was 10—36 days; for peptic ulcer, 6—26 days; for hernia repair, 2—12 days; for lens extraction, 4—18 days; for excision of the semilunar cartilage, 3—21 days; and for hysterectomy, 3—18 days [13]. Similar variation has been shown in studies conducted for various states in the United States and among HMOs [41].

In a recent review of the literature on misutilization in hospital care, Anthony L. Komaroff cites studies that show that between 5 and 20 percent of hospital days have been judged to be inappropriate ([15], p. 237). The findings are corroborated in another review by P. M. Gertman and J. D. Restuccia whose own recent work suggests that over 25 percent of days of care represent inappropriate use of the hospital ([11], pp. 863—864). Finally, Harold Luft's summarization of HMO studies provides support for the conclusion that large amounts of unnecessary care are provided by demonstrating that where economic incentives exist the utilization of hospital days can be reduced 20 to 40 percent with no discernible effect on the quality of care [19].

Annual surgery rates in different hospital catchment areas serving apparently homogeneous populations in the same state have been found to vary from a low of 360 per 10,000 population to a high of 689 per 10,000 [43]. John Wennberg notes that $10 million could have been saved over a ten-year period by eliminating the 2500 extra hysterectomies done in one high-rate market area in Maine compared to another low-rate area ([41], p. 15). Other studies have documented similar variations elsewhere [39]. Studies on inappropriate utilization of surgical services were recently reviewed by E. G. McCarthy, *et al.* [20]. McCarthy's own study of a mandatory second opinion surgery program showed that 17 percent of elective procedures were not confirmed for surgery by the consultant, therefore representing potentially unjustified services ([20], p. 13).

The use of diagnostic tests and procedures has been shown to have a threefold variance among physicians whose practices are similar in terms of diagnostic mix and caseload [33]. Excessive use of laboratory tests and x-rays has been suggested by a number of other studies which show that many procedures are performed which have little or no effect on the outcome of care or the treatment process [15], [32]. Komaroff's review of this literature showed that only a small percentage of laboratory tests ordered were actually used in formulating diagnostic and therapeutic decisions ([15], p. 243).

Non-medical factors which have been correlated with physician decisions have been classified by Rothert et al., as demand factors such as the extent of insurance coverage and the patient's desire for treatment, supply factors such as the availability of hospital beds and the number of physicians per capita, and organizational factors such as whether the physician practices in a group [31]. These factors have been shown to influence, and in some cases to be decisive considerations, in physicians' recommendations for follow-up visits, elective surgical procedures, lab tests, x-rays, and hospital admissions [32]. This does not mean that these factors are the primary motivators of physician behavior. On the contrary, it has been argued by Wennberg that the importance of non-medical factors in explaining utilization simply reflects the uncertainty that exists in medical decision making: ". . . the enabling variable I think most important is the absence of objective standards or agreement on standards concerning the efficiency with which medical needs are met . . ." ([42], p. 517).

The empirical literature on utilization of medical care suggests that what is in the best interest of a patient with a given diagnosis or state of illness is far from settled. There appear to be significant differences in the kinds and degree of utilization of medical services for any given disease state and considerable evidence that some of this variation is the result of overutilization. Hence, it is problematic to assert that policies which encourage providers to give care at less cost will necessarily create ethical conflict. If, as the literature suggests, some significant portion of medical care currently provided is not necessary, then failing to provide such care cannot be properly called a violation of an ethical obligation to provide the utmost benefit to individual patients.

IV. DISTRIBUTIONAL CONCERNS

Although it may surprise some, medical care is already rationed and always has been. The poor and unemployed who cannot afford to pay for medical care in the United States either do not receive it or often accept inferior treatment. Another form of rationing concerns the allocation of such scarce resources as intensive care beds, dialysis machines, or donated organs, among the large number of patients who would benefit from them. People who live in rural areas often do not receive the same standard of care as those in urban areas simply because of time and transportation costs. The current controversy surrounding rationing seems more related to the fact that the costs of medical care are rapidly outstripping the ability or willingness of the middle class to pay than to a special sensitivity to the plight of the poor and unemployed. Yet, the plight of the latter groups pose significant social justice concerns which must be addressed under any rationing scheme. Unfortunately, the traditional approach tends to see the issue of rationing solely in terms of the conflict with physicians' obligations to particular patients; it overlooks the many individuals who do not receive medical services because of economic or other impediments to access.

An extreme case described by Duncan Newhauser and William B. Stason illustrates the distributional problems associated with the traditional view:

> A favorite example of cost-ineffective clinical decision making is that of Dr. Harvey Cushing at Ypres .... Cushing was an army surgeon during the war. Although the allied mortality was as much as 50,000 on some days, not counting the wounded, Cushing operated very carefully on only two patients a day ([25], p. 134).

In this example, where there was a serious shortage of resources, most people would argue that Dr. Cushing would have been justified in lowering his standard of care for his patients if this would have allowed him to more adequately serve the needs of the population of wounded soldiers. Indeed, most physicians would admit that such trade-offs often have to be made when allocating their limited time among the patients who need it, or deciding which patients should be admitted to the hospital's intensive care units. Unfortunately, the traditional view of medical ethics has little to say about such situations, because it abstracts from the existence of the scarcity of resources and tends to focus physician obligations on the immediate patient. If, however, scarcity is taken seriously, and with it the fact that some people in need of medical care will not receive it, then ethics would seem to require that considerations of justice be brought forward. The question at bottom concerns both the inherent conflicts between the provider's role as rationer and personal interest to restrain costs on the one hand and fundamental questions of social justice on the other. Without a coherent view of social justice that determines a distributional objective for medical services, however, the advantages and disadvantages of creating an explicit rationing role for physicians are certainly unclear. Lack of consensus on this point seems to be a serious impediment to settling the issue of physician obligations. Recent theoretical works, however, seem committed to providing a philosophical foundation for a distributional ethic of medical care.

For example, Norman Daniels offers just such a social justice criterion of a limited right to health care that may be helpful in dealing with this dilemma [7]. In answering the question: Do physician obligations to do all that is of benefit to patients conflict with obligations to distribute health care efficiently or equitably?, Daniels suggests that it depends on what entitlements patients have to an array of health care services, given their needs and limits on social resources. He asks:

> Does the patient have an entitlement — or even legitimate expectation — to every diagnostic test in the book to reduce the risk of some condition not being diagnosed properly? Or do demands of equity and efficiency place reasonable limits on what array of procedures are appropriate ones for the physician to consider? ([7], p. 133)

Daniels argues that justice considerations lead to limited entitlements. Relying on the relationship between equal opportunity and illness, he establishes a limited right to a basic minimum set of health services [8]. If we accept Daniels' argument for a limited entitlement, it follows that providers may be justified in denying beneficial services beyond that entitlement if this permits other patients to realize their rights. In similar fashion, the President's Commission for the Study of Ethical Problems in Medicine and Biomedical and Behavioral Research adopted a notion that society has a moral obligation to ensure equitable access to adequate care for all ([28], p. 4). As a result, physicians may practice rationing on the grounds that it ensures that the obligation will be met successfully.

With a coherent view of social justice that distinguishes between types of health care needs and services to which individuals are entitled, it might be possible to gain a consensus regarding the physician's role in cost control. On this basis it would be possible to argue that the obligation of the physician is to provide the best possible medical care that is *rightly* owed to patients under a basic minimum concept of care. This would require that beneficial care that is ultimately not worth the cost be treated as ethically permissible, but not ethically required as a matter of justice.[6] Those physicians who continue to believe that they have obligations to provide the utmost care to patients irrespective of cost could still provide such care *outside* the socially guaranteed basic minimum. Physicians, thus, would be free to practice according to their ethical convictions so long as the costs associated with such a medical practice were not unjustly shifted to society, or did not prevent others from realizing their right to a basic minimum. Key to the concept of the basic minimum is that the services which comprise it will be generous enough and yet affordable to assure a consensus.[7]

V. INFORMED CONSENT AND STANDARDS OF CARE

If there were a consensus regarding a basic standard of care to which all were entitled, it might be possible to draw a distinction between services to which one is entitled, those services which are basic and essential, and services which are potentially beneficial but are ultimately not worth the cost. Without such a consensus, however, a serious ethical conflict arises when providers are placed in the position of having to deny care that has a clear, but costly, benefit.

This conflict has been articulated in terms of the threat to the fiduciary relationship between doctors and patients. Luft, for example, has noted that: "Unethical behavior may arise from the loss of the physician as an impartial and trusted agent for the patient" ([19], p. 336). Similarly, David Mechanic has observed:

> In such cases as medical care, where the activity or production and the product are the same, trust has crucial significance. Much of the success of medical care is likely to depend as much on the consequences of trust (confidence and effective communication) as on technical virtuosity. I do not imply that trust in doctor-patient relationships is presently optimal; but I doubt that more competition would be an effective means to increase it ([21], p. 35).

These concerns, however, are hardly theoretical. They reflect a long-standing view of the practice of medicine as a fabric of rights and responsibilities which have a professional and legal articulation in the United States, an important example of which is malpractice law.

Currently, standards of care in malpractice proceedings are set without regard to questions of cost. James Blumstein has noted that medical malpractice is designed to achieve the objectives of quality assurance and victim compensation [4]. Quality assurance is promoted by focusing on input and process standards of medical care delivery. For this reason, a malpractice action focuses on the physician's training, facilities of the institution, and process-oriented aspects of quality assurance such as record keeping, use and interpretation of laboratory and other diagnostic tests, and provision of customary or accepted treatments ([4], p. 390). Since outcome measures of quality are deemphasized in malpractice litigation,

> a provider who seeks to ration medical care resources — for example, by ordering fewer or less costly tests, by providing a smaller margin of safety in terms of facilities or equipment availability, or by allowing less highly credentialed personnel to perform certain procedures — runs the risk of increased exposure to malpractice liability in the case of a medical maloccurrence ([4], p. 390).

A defense against a malpractice suit based on the practitioner's desire to save resources currently has little chance of succeeding. For this reason, proponents of efficiency will have to address the need to legislate standards of care based on considerations of cost if physicians are to be relieved of the risk of being sued for malpractice under standards based primarily on inputs.

One apparent way around this problem involves focusing on informed consent. Informed consent is based on the principle of self-determination. Practitioners cannot legally treat patients without obtaining their consent, and the consent must be "informed" for it to be valid. Traditionally, the information that practitioners disclose in pursuing informed consent is based on what a prudent physician would disclose in similar circumstances. However, in many states a broader definition has been adopted which focuses on what a reasonable patient in a given situation would want to know in order to make the decision [34]. These standards differ in terms of the placement of decision-making authority in medicine. The latter represents the standard adopted in 1972 in *Canterbury v. Spence*, a leading American case [5]. This standard is patient-oriented: it looks to the plaintiff's need or desire to know. The traditional rule is physician-oriented: it looks to the conduct of the reasonable physician ([34], p. 20).

A system of equitable allocation of limited resources may not be able to tolerate the financial effects of all reasonable possible patient decisions — the patient-oriented doctrine of informed consent. The problem this ideal creates concerns what the patient should be told when a beneficial option is not available because it is too costly. The right to know ethic suggests that a physician is obliged to inform a patient of all alternative treatments, because the judgment about which choice will best serve well-being properly belongs to the patient. This constraint would not apply, however, in cases in which the patient voluntarily consents to seek care from physicians who are known to consider costs in offering treatment options. The fact that people are willing to enroll in HMOs and PPOs in which physicians are known to limit the range of available alternative therapies illustrates this preference. If a particular physician knew that his patient thought him to be making economic as well as medical judgments in recommending treatment modalities, that understanding would become a part of the implied contract between them. By joining the HMO or requesting that his physician consider costs in offering different options, the patient consents in advance to limitations on choice.[8]

In addition, if there were a social consensus regarding a basic list of services to which all are entitled, then a line could be drawn whereby the physician would be justified in not telling the patient of something that could be done, but was not worth doing in the societal context. To not so inform would not violate the patient's right to know since that right would be restricted to *available* care as defined by the basic minimum. Lacking a definition of a decent minimum and/or prior consent, however, the informed consent doctrine as a right to be told of *all,* even marginally, beneficial care may pose a serious constraint on a provider's ability to ration. The traditional view of the agency role of the physician seems to obligate the physician to do what the informed patient would choose to have done. On the face of it, any modification of standards of care or informed consent will have serious ramifications on the legal status of health care in the United States; correlatively, any change in the status of health care will entail difficult policy choices regarding malpractice law.

VI. CONCLUSION

In this paper I discussed how the application of efficiency concerns to patient care decision conflicts with the traditional view of medical ethics. I also pointed out, however, that uncertainty regarding the benefits of medical care, concerns about the appropriateness of the traditional view of medical ethics, and problems associated with determining standards of care and informed consent requirements make it difficult to assess the implications of policies which encourage providers to adopt efficiency concerns. These complexities suggest an agenda for future research and discussion.

In order to better evaluate current policy trends it is necessary to know more about what medical procedures are cost-effective and about the amount of unnecessary care in the current system. Only then will it be possible to address adequately the ethical problem posed when contemplating a particular policy decision.

In order to achieve a fairer distribution of limited resources, the traditional view of medical ethics may have to be broadened to recognize some notion of social obligation regarding access. What is needed is a social ethic that helps doctors accept the rationing role by defining a minimal set of services that society owes to all patients. The medical profession now has norms concerning what constitutes bad medical practice. Those norms may have to be expanded to include cases in which high costs are not justified by marginal expected benefits. That, in turn, requires a consensus on the basic minimum which is hardly an easy task.

Finally, if physicians are to accept a rationing role, then accommodations in standards of care and informed consent requirements will have to be made. Patients will have to accept the fact that certain options which may be beneficial to them will be withheld and standards of care will have to be defined in a way that recognizes resource limitations.

School of Public Health
University of Texas Health Science Center
Houston, Texas

NOTES

[1] Such efforts included: government review of the construction and equipment acquisition decisions of hospitals and nursing homes (Certificate Of Need); utilization reviews of medical records by physician committees (Professional Standards Review Organizations); physician review of surgery decisions (second-opinion programs); federally-funded planning committees (Health Systems Agencies); and limits on the allowable revenues that hospitals could receive (rate review). For a detailed review of these programs and their effects on health care costs, see ([14], [37], [46]).

[2] This approach has been adopted in large part because the regulatory attempts to restrain cost increases proved largely ineffective. See [16].
[3] For a more detailed review of the cost control incentives created by these policies, see ([21], [22]).
[4] See ([3], [19], [26]) for discussion of ethical problems related to the issues of cream skimming, health care vouchers, and HMO practice.
[5] This so-called "traditional view" of medical ethics is discussed in greater detail in [22] and criticized in the paper by Agich elsewhere in this volume.
[6] The equality of opportunity principle is compatible with a two-tiered system of distribution. The basic tier would include those services that meet valid health care needs determined by their importance in contributing to equality of opportunity. The upper tier should not be covered since the benefits of these services are non-essential.
[7] For a discussion of the conditions which must be met to achieve the basic minimum, see the paper by Baily elsewhere in this volume.
[8] For a general discussion of the problem of rationing and publicity, see the paper by Winslow elsewhere in this volume.

BIBLIOGRAPHY

[1] Agich, G. J. and Begley, C. E.: 1985, 'Some Problems with Pro-Competition Reforms', *Social Science and Medicine 21,* 623—630.

[2] Angell, M.: 1985, 'Cost Containment and the Physician', *Journal of the American Medical Association 254,* 1203—1207.

[3] Arras, J. D.: 1982, 'Health Care Vouchers and the Rhetoric of Equity', *Hastings Center Report 11,* 29—39.

[4] Blumstein, J.: 1983, 'Rationing Medical Resources: A Constitutional, Legal, and Policy Analysis', in President's Commission for the Study of Ethical Problems in Medicine and Biomedical and Behavioral Research, *Securing Access to Health Care, Volume Three: Appendices: Empirical, Legal, and Conceptual Studies,* U.S. Government Publishing Office, Washington, D.C., pp. 349—394.

[5] Canterbury v. Spence: 1972, 464 F. (2d) 772.

[6] Daniels, M. and Schroeder, S. A.: 1977, 'Variation Among Physicians in Use of Laboratory Tests. II. Relation to Clinical Productivity and Outcomes of Care', *Medical Care 15,* 482—487.

[7] Daniels, N.: 1981, 'What is the Obligation of the Medical Profession in the Distribution of Health Care?' *Social Science and Medicine 15F,* 129—133.

[8] Daniels, N.: 1982, 'Equity of Access to Health Care: Some Conceptual and Ethical Issues', *Health and Society 60,* 51—81.

[9] Donabedian, A.: 1979, 'The Quality of Medical Care: A Concept in Search of a Definition', *Journal of Family Practice 9,* 277—284.

[10] Freeland, M. A. and Schendler, C. E.: 1984, 'Health Spending in the 1980's: Integration of Clinical Practice Patterns with Management', *Health Care Financing Review 5,* 1—68.

[11] Gertman, P. M. and Restuccia, J. D.: 1981, 'The Appropriateness Evaluation Protocol: A Technique for Assessing Unnecessary Days of Hospital Care', *Medical Care 19,* 855—871.

[12] Gibson, R., Waldo, D. R. and Levit, K. R.: 1984, 'National Health Expenditures, 1983', *Health Care Financing Review 6*, 1—54.

[13] Heasman, M. A. and Carstairs, V.: 1971, 'Inpatient Management: Variations in Some Aspects of Practice in Scotland', *British Medical Journal 1*: 495—498.

[14] Joskow, P. L.: 1981, *Controlling Hospital Costs*, MIT Press, Cambridge, Massachusetts.

[15] Komaroff, A. L.: 1983, 'The Doctor, The Hospital, and the Definition of Proper Medical Practice', in President's Commission for the Study of Ethical Problems in Medicine and Biomedical and Behavior Research, *Securing Access to Health Care, Volume Three: Appendices: Empirical, Legal, and Conceptual Studies*, U.S. Government Printing Office, Washington, D.C., pp. 225—251.

[16] Lairson, D. R.: 1985, 'Prelude to Prospective Reimbursement by DRGs', *Journal of Health and Human Resources Administration 8*, 6—18.

[17] Leaf, A.: 1984, 'The Doctor's Dilemma And Society's Too', *New England Journal of Medicine 310*, 718—721.

[18] Levinsky, N. G.: 1984, 'The Doctor's Master', *New England Journal of Medicine 311*, 1573—1575.

[19] Luft, H.: 1983, 'Health Maintenance Organizations and the Rationing of Medical Care', in President's Commission for the Study of Ethical Problems in Medicine and Biomedical and Behavioral Research, *Securing Access to Health Care, Volume Three: Appendices: Empirical, Legal, and Conceptual Studies*, U.S. Government Printing Office, Washington, D.C., pp. 313—315.

[20] McCarthy, E. G., Finkel, M. L., and Ruchlin, H. S.: 1981, *Second Opinion Elective Surgery*, Auburn House, Boston.

[21] Mechanic, D.: 1976, 'Rationing Health Care: Public Policy and the Medical Marketplace', *Hastings Center Report 6*, 34—37.

[22] Morreim, E. H.: 1985, 'Cost Containment: Issues of Moral Conflict and Justice for Physicians', *Theoretical Medicine 6*, 257—279.

[23] Newhauser, D.: 1976, 'The Really Effective Health Service Delivery System', *Health Care Management Review 1*, 25—32.

[24] Newhauser, D. and Lewicki, A. M.: 1973, 'What Do We Gain From the Sixth Stool Guaiac?' *New England Journal of Medicine 293*, 226—228.

[25] Newhauser, D. and Stason, W. B.: 1979, 'Cost-Effective Clinical Decision Making', in E. J. Carels, *et al.* (eds.), *The Physician and Cost Control*, Olegeschlager, Gunn & Hain, Inc., Cambridge, Massachusetts, pp. 133—149.

[26] Newhouse, J. P.: 1982, 'Is Competition the Answer?', *Journal of Health Economics 1*, 110—116.

[27] Pellegrino, E. O.: 1978, 'Medical Morality and Medical Economics', *Hastings Center Report 9*, 8—11.

[28] President's Commission for the Study of Ethical Problems in Medicine and Biomedical and Behavioral Research: 1983, *Securing Access to Health Care, Vol. 1: Report*, U.S. Government Printing Office, Washington, D.C.

[29] Relman, A. S.: 1985, 'Cost Control, Doctors' Ethics, and Patient Care', *Issues in Science and Technology 1*, 103—111.

[30] Roos, N. P.: 1984, 'Hysterectomy: Variation in Rates Across Small Areas and Across Physicians' Practices', *American Journal of Public Health 71*, 606—612.

[31] Rothert, M. L., *et al.*: 1984, 'Differences in Medical Referral Decisions', *Medical Care 22*, 42—50.

[32] Schroeder, S. A.: 1979, 'Variations in Physician Practice Patterns: A Review of Medical Cost Implications', in E. J. Carels, *et al.* (eds.), *The Physician and Cost Control,* Olegeschlager, Gunn, & Hain, Cambridge, Massachusetts, pp. 23—50.

[33] Schroeder, S. A., *et al.*: 1973, 'Use of Laboratory Tests and Pharmaceuticals: Variation Among Physicians and Effect of Cost Audit on Subsequent Use', *Journal of the American Medical Association 225,* 969—973.

[34] Schwartz, R. and Grubb, A.: 1985, 'Why Britain Can't Afford Informed Consent', *Hastings Center Report 15,* 19—25.

[35] Schwartz, W. B. and Joskow, P. L.: 1978, 'Medical Efficacy Versus Economic Efficiency: A Conflict in Values', *New England Journal of Medicine 299,* 1463.

[36] Speedling, E. J. and Rose, D. N.: 1985, 'Building an Effective Doctor-Patient Relationship: From Patient Satisfaction to Patient Participation', *Social Science and Medicine 21,* 115—120.

[37] Steinwald, B. and Sloan, F. A.: 1983, 'Regulating Approaches to Hospital Cost Containment: A Synthesis of the Empirical Evidence', in M. Olson (ed.), *A New Approach to the Economics of Health Care,* American Enterprise Institute, Washington, D.C., pp. 273—308.

[38] Thurow, L. C.: 1985, 'Medicine Versus Economics', *New England Journal of Economics 313,* 611—614.

[39] Vayda, E.: 1973, 'A Comparison of Surgical Rates in Canada and in England and Wales', *New England Journal of Medicine 289,* 1224—1229.

[40] Weinstein, M. C. and Stason, W. B.: 1977, 'Foundations of Cost-Effectiveness Analysis for Health and Medical Practice', *New England Journal of Medicine 296,* 716—721.

[41] Wennberg, J. E.: 1984, 'Dealing With Medical Practice Variations: A Proposal for Action', *Health Affairs 3,* 6—32.

[42] Wennberg, J. E.: 1985, 'On Patient Need. Equity, Supplier-Induced Demand, and the Need to Assess the Outcome of Common Medical Practices', *Medical Care 23,* 512—520.

[43] Wennberg, J. E. and Gittelsohn, A.: 1975, 'Health Care Delivery in Maine, I: Patterns of Use of Common Surgical Procedures', *Journal of the Maine Medical Association 66,* 123—149.

[44] Wennberg, J. E. and Gittelsohn, A.: 1982, 'Variations in Medical Care Among Small Areas', *Scientific American 246,* 120—134.

[45] West, R. R. and Carey, M. J.: 1978, 'Variation in Rates of Hospital Admission for Appendicitis in Wales', *British Medical Journal 1,* 1662—1664.

[46] Zeckhauser, R. and Zook, C.: 1981, 'Failure to Control Health Costs: Departures from First Principles', in M. Olson (ed.), *A New Approach to the Economics of Health Care,* American Enterprise Institute for Public Policy Research, Washington and London, pp. 87—128.

MARC D. HILLER AND ROBIN D. GORSKY

SHIFTING PRIORITIES AND VALUES: A CHALLENGE TO THE HOSPITAL'S MISSION*

What should be the driving force underlying the health care industry in the United States as the twenty-first century approaches? Should it be to maximize the quality of health care and availability of resources for the population or to maximize the return on investment through pricing and service policies? This dichotomy poses what possibly may be the most complex and far-reaching issue ever to confront the American health care system. Thus far, most of the debate on this issue has centered either at the societal (macro) level where questions of access to medical care are addressed, or on the individual patient care (micro) level, involving obligations within the context of the provider-patient relationship. Often overlooked in such discussions has been the institutional, or meso, level ([17], pp. 7—9; [18]). This level concerns the responsibilities of those who manage health care institutions, namely administrators. The classic dilemma arising at this level is the conflict between the traditional charitable healing mission of the health care institution, which is oriented toward benefiting patients without recognizing a limit to resources, and the so-called business ethic, which is geared toward maximizing institutional solvency and, in a growing number of cases, profitability [8].

The objective of this paper is to explore how the factors which have influenced the mission of the health care institution also have affected the values of administrators. First, the concept of the institution as a moral agent is introduced. Second, the historical evolution of the community hospital in America is traced from its origins in the 1700s. Included in this discussion is an account of the variety of developments which have influenced the hospital's mission and have been responsible for the major changes which have occurred. Third, contemporary trends which are affecting the status and overall orientation of hospitals in the late twentieth century are described and the implications of these changes are discussed in terms of the ethical dilemmas and value conflicts confronting those who manage contemporary community hospitals.

G. J. Agich and C. E. Begley (eds.), The Price of Health, 245–261.

I. INSTITUTIONS AND ADMINISTRATORS AS MORAL AGENTS

Philosophers such as Edward Pellegrino, David Thomasma, and Richard DeGeorge have argued that the overall health care institution and those responsible for its operation bear a collective ethical obligation as so-called "moral agents" ([9], [29], [40]). According to Thomasma, "all hospital staff members are moral agents, as is the institution itself. Administrators . . . should not restrict their role to attending to bureaucratic concerns, but should insist on the moral character of the institution" ([40], p. 78). He argues that institutions have moral obligations to their patients, their patients' families, their staff members, their community, as well as to their own survival. Yet, while working to ensure the latter, Pellegrino insists that hospitals' foremost obligation is to their patients, regardless of such institutional concerns as costs or profits [31].

In recognizing the paramount obligation of the institution to the patient, Pellegrino and Thomasma argue:

> A hospital, by the very fact of its existence in a community, makes a declaration; that is, it professes to concentrate and make available those resources which a person can call upon if he is ill. Implicit in that profession is a promise to assist the sick person to regain what he has lost — his health — at least to the maximum degree possible ([30], pp. 250–251).

Thus, Pellegrino and Thomasma equate institutional morality to that morality which guides practitioners in serving individual patients. While health care institutions obviously are not human beings, the collective moral responsibility of the institution can be assumed by those within it ([9], p. 90). Whether this collective responsibility may be centered on a single chief executive officer, held jointly by several persons as in some form of executive committee, or distributed throughout the organization as in a decentralized management model, it rests with administration.

One might debate whether the hospital or its administrators do or do not constitute "third parties" in the context of health care delivery. If the hospital and its administration are viewed simply as an extension of the physician, then they are bound logically by the same ethical obligations. If considered separate, then arguments concerning the moral agency role of institutions and their administrators should be applicable. The latter perspective is assumed here, suggesting the importance of the institutional mission as distinct from the ethic domination the physician-patient relationship.

Little argument exists that both hospitals and health administrators traditionally have espoused a commitment to a "healing" mission similar to that of physicians and other providers. Despite conflicts at the meso level between the institutional mission, namely that of the hospital as a moral agent, and the administrative ethic representing the professional standards and values of health administrators, both have emphasized patient beneficence as a priority [18].

Following considerable effort to develop a professional code of ethics, the American College of Healthcare Executives (formerly the American College of Hospital Administrators) decided that an administrator's code logically could not be separated from a code applying to all hospital personnel. Thus, collaboration between the American College of Healthcare Executives (ACHE) and the American Hospital Association (AHA) produced a joint ethical code, entitled "Hospital Ethics", which was ratified by the governing bodies of both organizations. It sets forth an ethical mission for hospital personnel that embraced charity and caring as central values [25]. According to the original 1941 Code:

> The hospital is to render care to the sick and injured as its primary responsibility. Financial concerns and other interests should be of secondary consideration. The duty of the hospital is also to advance scientific knowledge and education of all participating in the work, and to take an active part in the promotion of general health [2].

By setting financial and other considerations as secondary, the original Hospital Code of Ethics stressed both the traditional healing and charitable missions of the institution. Hospital administrators, however, have competing obligations and interests [3]. Whereas the profession of medicine historically has deemed patient best interest as sovereign, the large scale and complexity of contemporary hospitals demands varying degrees of commitment to other priorities as well, priorities which, Beauchamp and McCullough point out, "tend to cluster around considerations of efficiency, with special emphasis on cost-effectiveness and cost-reduction, and, increasingly, profit" ([4], p. 145).

Changes in health care have inevitably resulted in changes in the Code. With the revision of the joint ACHE/AHA Code of Ethics in 1957, the emphasis on the healing mission of the institution remained while the charity directive became less emphatic:

> Recognizing that the care of the sick is their first responsibility and a sacred trust, hospitals must at all times strive to provide the best possible care and treatment to all in need of health care. Such institutions, cognizant of their unique role of safeguarding the nation's health, should seek through compassionate and scientific care and health education to extend life, alleviate suffering, and improve the general health of the community they serve [2].

In part the AHA's unwillingness to adjudicate ethics violations, as well as failing to recognize the distinction between ethical obligations of the administrator and those of the institution, led to a separation of the joint ethical code in 1967. The AHA revised its Code in 1967 under the title "Guidelines on Ethical Conduct and Relationships for Health Care Institutions" and further revised them in 1974 and 1980. Examples of changes made in the AHA Guidelines include modifications of earlier provisions that disallowed institutional marketing of services or solicitation for patients. Further changes resulted in acceptance of competition among hospitals in the late 1970s ([25], p. 76).

In turn, the ACHE developed its independent "Code of Ethics", including provisions relating to the accountabilities of administrators to the College and to their actions as institutional executives. It was subsequently revised in 1973 and 1983. As of this writing, the ACHE is in the midst of a major revision of its ethical code during 1986.

While hospital administrators remain ethically bound to adhere to the ACHE Code of Ethics, they are also expected to adhere to the AHA's guidelines for health care institutions. Such an expectation carries on the 1941 dual commitment toward self-conduct as professional health care administrators and institutional ethics. As stated in the Preamble of the ACHE Code of Ethics:

> The ACHE Code of Ethics is specifically designed to define and set forth broad guidelines applicable to the personal (professional) accountability of all members of top-level management involved in the administration of hospitals, other health institutions, and related health activities represented in the mix of its membership. It is united in purpose with the Code of Ethics of the American Hospital Association, which defines standards of conduct for health institutions [1].

The most recent ACHE Code acknowledges the influence of economic pressures and the scarcity of resources. In maintaining the executive's institutional accountability, the introductory provision states that one must "recognize that the care of the ill and injured is a prime responsibility and at all times strive to provide all in need of health services quality care and treatment *consistent with available resources*" [emphasis added] [1].

As observed in the following review of the actual history of the development of community hospitals in America, the healing mission of the hospital generally has not been questioned. However, the institution's ability to maintain a charitable orientation appears recently to have been placed in question. Moreover, to the extent that the healing mission is inextricably linked to providing services to those in need regardless of ability to pay ([7], [8], [16], [29], [30], [31], [32], [33], [41], [42]), this mission too, may be subject to change amid the current economic era. A longstanding analyst of trends in American hospital, Robert M. Cunningham, Jr. has warned that hospitals may be forgetting their healing mission in response to the demand for a business orientation and concern for profits [8].

II. DEVELOPMENTAL PERSPECTIVE ON THE HOSPITAL MISSION

The history of hospitals is traceable to the almshouses and pesthouses that emerged in many American communities by the mid-1700s [24]. Both almshouses and pesthouses were maintained for the poor and the homeless, and were avoided by most others. These institutions offered poor sanitary conditions, were crowded, poorly heated and ventilated, and staffed with poorly trained nursing and medical personnel ([10], pp. 173–174).

Almshouses, also referred to as poorhouses or workhouses, were established by city governments to provide food and shelter to the homeless poor. Only secondarily did they offer medical care. According to B. J. Stern, the earliest hospitals, if they could be called such, were the poorhouses such as Henricopolis in Virginia (1612), Blockley in Philadelphia (1732), Charity Hospital in New Orleans (1736), and the Public Workhouse and House of Correction in New York City (1736) [39]. Not until the late 1800s did the hospital units within almshouses evolve into separate medical care institutions, constituting the first public hospitals ([10], p. 173). Later, many almshouses became tuberculosis hospitals.

The first government institution designed solely for the care of the sick was the "pesthouse" built on the same grounds as the New York workhouse ([12], pp. 28–29). Pesthouses were local government facilities established as quarantine units in seaports in order to contain the spread of infectious diseases. As in almshouses, medical care was a secondary function to controlling the spread of infectious diseases.

The first community-owned or voluntary, non-profit hospitals developed in the late 1700s. Among the earliest voluntary hospitals were Pennsylvania Hospital, Philadelphia (1751), New York Hospital, New York City (1773), and Massachusetts General Hospital, Boston (1816) ([10], p. 174). Except in major cities in which the concentration of poor was too large, voluntary hospitals typically were community focused and served those in need of medical care who were unable to pay on a charitable basis. These institutions were supported through community contributions and private philanthropy. Members of the medical staff donated their time ([24], [34]). Services in these hospitals focused on treatment of accidents and acute illnesses excluding care for contagious diseases and mental illness which were viewed as a governmental responsibility.

Many city, county, and state governments established state mental hospitals and public (for example, city or municipal) hospitals between 1770 and 1850, since voluntary hospitals specifically avoided providing treatment for contagious diseases and mental illness. This time period also marked the advent of hospitals established by various religious orders.

While the care in the early voluntary hospitals was notably better than that provided by the poorhouses which preceded them, it was not until the late 1800s that voluntary hospitals became the preferred source of treatment of the more affluent members of society. Until then, most of the more affluent preferred to be treated at home.

Embodied within the mission statements of the early voluntary hospitals was the objective to provide health care to anyone in need. As Benjamin Franklin stated in recounting the history of the Pennsylvania Hospital: "the purpose of the hospital is to provide the means whereby the sick could be treated and restored to health and comfort and become useful to themselves, their families, and the public for years thereafter. . . . The religious motivation was subordinate, but not altogether absent" ([8], p. 88).

Rather significant changes occurred between the mid-1800s and the mid-1900s that significantly influenced the evolution of hospitals. Between the late 1800s and the first decades of the twentieth century, the health care system experienced major scientific advancements in medicine which unquestionably contributed to hospital progress. The most significant advances included the discovery of anesthesia and the rapid progress in surgery that followed, the development of the germ theory of disease, and the subsequent advances in aseptic and sterilization techniques. Collectively, these advances significantly contributed to transforming the hospital's custodial role of isolating and sheltering the poor to an institution professing a therapeutic mission and receiving broad community support ([6], [34]). By 1900, 40 percent of hospitalizations were for surgery [10].

The first hospital laboratory opened in 1889, and x-ray film was first used in medical diagnosis in 1896. These developments marked milestones in the history of medical diagnosis; subsequent technological developments in the medical laboratory continued to increase the role of the hospital as a diagnostic center as well as a therapeutic center.

The discovery of insulin in 1923, liver extract in 1929, sulfonamides in the mid-1930s, coupled with the widespread use of antibiotics beginning in the early 1940s, shifted the role of the hospital toward caring for those afflicted with chronic diseases such as heart disease, cancer, and stroke.

The 1910 Flexner Report brought dramatic changes to medical education and also had a major impact on hospitals [11]. Among the changes brought about as a result of this report was a strong emphasis on the scientific basis of medicine, the establishment of standards of medical education, and the enactment laws relating to accreditation and licensure. University-based medical education became standard and clinical training was centered in hospital wards. The hospital rapidly evolved as a hub for medical education and research in addition to patient care. Requirements for medical education forced an expansion of hospital facilities, services, equipment and personnel. Further, as medicine became more specialized in the 1920s and 1930s, a proliferation of hospital-based internships and residencies emerged as a cornerstone of advanced medical education. With the rapid growth of basic medical sciences and education, physicians' use of hospitals as their workshops led to the increasing importance and centrality of the hospital ([37], p. 178).

III. DEVELOPMENTS IN INSURANCE AND HEALTH CARE FINANCING

As the modern hospital developed and its benefits became more widely recognized and utilized, a broader and more stable source of financial support was required. The Great Depression demonstrated the inability of local governments and voluntary efforts to sufficiently meet the resource needs of hospitals; many patients were unable to pay their bills and hospital use declined. The response to the problem was the development of health insurance, which offered consumer protection against the risk of hospital expenses, while insuring hospitals both utilization and timely payment for services rendered.

The influence of health insurance on hospitals is noteworthy. Beyond ensuring the financial stability of the institution, it provided the necessary funds to facilitate a considerable expansion of facilities and services, which promoted the development and use of new medical technology.

After World War II another source of available resources for health care was the federal government. Direct federal involvement in the hospital industry was marked by enactment of the Hill-Burton (Hospital Survey Construction Act) begun in 1946. By 1974, when Hill-Burton was replaced by the National Health Planning and Resources Development Act, Hill-Burton had been responsible for the construction of nearly 40 percent of the acute care hospital beds in the country. Other significant initiatives were launched to support medical research and education.

From assisting with the development of health care resources, the federal role in the hospital industry was expanded to include financing the delivery of health care for large segments of society, specifically the poor, aged, and disabled. With the establishment of Medicare and Medicaid in 1965 to assist these groups, government at all levels became a major payor of medical care. By 1984, approximately 54 percent of the nation's health care expenditures were made by government, mostly through Medicare and Medicaid.

With increasing revenues becoming available through health insurance and government programs, the stage was set for profitmaking activity in health care and the rise of the investor-owned hospital. Through the 1960s and 1970s, marketplace economics and increasingly sophisticated technology were growing and an increased commercialization of health care became evident. Fostered by the highly efficient business practices dominating the proprietary sector, the hospital industry as a whole began to champion management practices directed at enhancing institutional efficiency and strengthening financial position.

Into the 1980s, the dominant presence of the large investor-owned hospital chains has continued to foster further corporatization and privatization, even among voluntary hospitals. Corporate restructuring and reorganization, vertical and horizontal integration, for-profit subsidiaries of non-profit institutions, and policies directed at maximizing revenue generation promote similar practices among hospitals regardless of ownership status.

While the present dimensions of the for-profit sector in health care are unknown, crude estimates of gross revenues of the investor-owned health care sector were in excess of $40 billion in 1980 [30]. In 1984, nearly 30 percent of multihospital systems acute care beds were controlled by four investor-owned corporations (Hospital Corporation of America, Humana, American Medical International, and National Medical Enterprises). In contrast, the four leading non-profit hospital chains (Adventist Health Systems/US, New York City Health and Hospitals Corporation, Kaiser Foundation Hospitals, and Sisters of Mercy Health Corporation) controlled only about 10 percent of total beds ([22], p. 91). Collectively, in 1984, investor-owned chains controlled 12.5 percent of all non-federal beds in the United States; their market share was about 8 percent and increasing ([35], pp. 3—4).

For-profit organizations are even more predominant among health care facilities other than hospitals. More than 75 percent of nursing homes are proprietary [5] and about 40 percent of the hemodialysis in the United States are provided by proprietary organizations [32]. Such organizations presently provide emergency medical services, home care, mobile CAT scanning, cardiopulmonary testing, industrial health screening, rehabilitation services (including sports medicine programs), dental care, weight control programs, comprehensive prepaid HMO programs, and laboratory and related services ([16], p. 3).

Thus, private investors have become a dominant force within the institutional sector of health care, and the profit motive has become a significant part of the hospital mission [13]. The conflict between this objective, based on maximizing the return-on-investment (ROI) to stockholders and/or striving for improved financial position to ensure institutional survival, and the traditional charity and healing mission, based on the ethical principle of beneficence, is rather obvious. Often it forces choice between what is in the best interest of the institution (or its investors) and what is in the best interest of individual patients to whom the institution voices commitment.

There are two business strategies noted by Miles F. Shore and Harry

Levinson which investor-owned hospitals may follow [36]. The first strategy focuses on long-term continuity and growth, by establishing a distinctive organizational character, quality, and type of service. The second strategy is to organize around short-term profits and expediency with the bottom-line as the measure of success. If the first strategy is followed, there may be little difference between profit making behavior and that of non-profit institutions. In the long run, this strategy may maximize ROI while fulfilling other more traditional goals as well. The survival of the traditional mission of the hospital that places beneficence above short-term profits will depend on which view is taken by the investors. According to Shore and Levinson:

> ... it seems unlikely that hospitals can go very far in becoming businesslike in the expedient sense without destroying their credibility with society, their patients, and their staffs. To exploit patients and staff by focusing only on business efficiency and profit, without concern for the organizational character of the hospital as a place that cares for the sick and that maintains high professional standards, spells institutional demise ([36], p. 320).

IV. ETHICAL IMPLICATIONS OF GROWTH IN THE FOR-PROFIT SECTOR

The major ethical issues associated with these changes in health care concern questions of access, quality of care, and cost-containment. While it is too early to offer empirical evidence, a variety of hypotheses have been suggested regarding the probable impact of for-profit medical practice on these system goals.

Generally speaking, negative consequences of for-profit medical practice are expected in terms of access and quality of care. The desire for revenues to exceed costs defines the profit motive. A health care institution receives revenue from various sources: patients, private insurance, government, private investors, and philanthropists. These revenues must exceed the costs of and resources needed to provide necessary services. One issue concerns what is done with profits. According to Kevin O'Rourke, investor-owned institutions siphon scarce dollars out of the health care system into the pockets of investors far-removed from the health care institution's mission ([28], p. 24). To the extent that profits are diverted from the hospital, then less revenues are available with which to buy resources to transform into services. Fewer resources may mean fewer services, which implies less access to services for large resource users, especially those who are unable to pay.

Another issue concerns ability to pay. Patients who can pay for health care either directly or through insurance will provide a source of revenue and will be welcome. Patients who provide no revenue may not be welcome since no other source of revenue can cover their expenses. If all must pay their own way, then any patient who cannot pay constitutes a drain on revenue and makes the institution that much less financially viable and that much less likely to survive. Further, for those investor-owned institutions which have been committed to return a specified percentage of profit on investments or to generate excessive dividends, corporate directives may limit the amount of "free care" that might be provided.

One strategy to enhance profits is not to admit patients whose financial resources are insufficient or whose resource use will be excessive — that is, to select patients on their ability to pay. Since the poor are generally higher-than-average resource utilizers, it is fiscally sound not to admit them.

Another strategy is to transfer non-revenue-enhancers to another institution or simply refuse patient admission in the absence of evidence of an ability to pay (a procedure termed "patient dumping"). The practice of transferring the indigent to a public institution certainly affects access. Several recent reports have provided graphic examples of the magnitude of the patient dumping problem and its effects on the poor and uninsured ([14], [15], [19], [20], [26]). Such practices portray the disappearance of the charity ethic in health care in favor of providing only for those who have the revenue source to cover the costs of the resources consumed in treatment.

In analyzing the investor-owned corporations and their impact on access, Paul Starr has noted that "profit-making enterprises are not interested in treating those who cannot pay. The voluntary [nonprofit] hospital may not treat the poor the same as the rich, but they do treat them and often treat them well" ([37], p. 448). The fear is that the pursuit of profits may promote a multi-tiered health care system. The first tier will be government-funded patients, paid for prospectively at the least amount that society feels it can collectively afford. Each patient will provide $X of revenue. The second tier will be employee-benefit coverage by private insurance, that is, health care negotiated by employers for their workers. Since these employers want to reduce cost, negotiations with hospitals and other health care providers will provide the greatest amount of services at the lowest possible price. Hospitals will trade increased volume (revenue) for decreased profits per person or diagnosis. Each patient in the second tier will provide $(X+Y).

The third tier will be paid for privately through coinsurance or personal resources. Within this tier, price is no object. Hospitals will receive what the market will bear: $(X+Y+Z). Hospitals whose major motive is profit maximization will seek out, court, and market to the third tier, to receive the extra $Z. Already, birthing centers and non-emergency ambulatory care facilities in hospitals are catering to the upwardly-mobile segment of the population who have either sufficient private resources or the coinsurance to buy "the best" and the most convenient medical care. This trend would reduce access for those not in the first or second tiers.

To the extent that the "survival" ethic or the profit drive predominates, risk of sacrifices in quality of care are real. Profit can be enhanced by increasing revenue or by decreasing costs. If patients pay less (or the hospital is reimbursed less per patient), the hospital's profits decrease. If patients use more resources per diagnosis or per inpatient episode, the hospital's profits decrease.

One strategy for lowering resource use under a fixed revenue scheme is to discharge patients early from the facility. This early discharge has direct implications for quality of care. Transferring patients to other hospitals has quality of care implications as well. The process of transferring a patient may in fact have adverse effects on the outcome of the diagnosis.

There is little question that as the overall costs associated with the health care system expand, the limit of available resources will become increasingly apparent — particularly to those least able to afford the price of health care. Further, the cost problem accentuates the institution's dilemma of attempting to provide care to everyone in need while having to maintain sufficient resources to ensure institutional survival. Thus, these are implications both for the institution, which is forced to bear rising costs in purchasing resources, and the patient, who is forced to pay higher costs to ensure access to quality care.

Studies have shown that there may be increased social costs associated with the growth of investor-owned institutions ([23], [27], [33], [37], [38]). This situation may occur as institutions practice "skimming" the third tier, namely those who are willing or able to pay more for services. This situation also may occur as the result of corporate reorganization, the purchasing of other institutions, or the diversion of revenue from hospitals to other investment opportunities. Regardless of the cause, when the institutions' costs rise, the prices they charge must also increase in order to maintain an attractive profit margin for investors.

Such increase leads to a variety of ethical dilemmas. As costs of health care continue to rise, fewer individuals will be able to afford access to needed services. And, in an effort to market or provide services at a more affordable price, some institutions may be willing to sacrifice quality. Given the prevailing trend for the poor to be driven to public hospitals often as a matter of desperation when access cannot be found elsewhere, the survivability of such institutions will be severely challenged. Plagued by a disproportionate number of non-paying patients, the rising costs associated with the provision of health care further jeopardizes their continued operation.

V. CONCLUSION

Unquestionably, the move toward increased efficiency through better management in health care is long overdue. To the extent that investor-owned health care institutions have prompted realization of this need, the entire health care industry – including the private, not-for-profit, and the public sectors – will benefit. Yet, in striving for a better bottom-line, institutional administrators must not forget the historic charitable healing mission that has dominated the delivery of health care and the health care professions for centuries. As O'Rourke has concluded, "assuring access to health care is a social obligation that must be borne by all the people and institutions that make up our society" ([28], p. 23).

Health administrators frequently find themselves on the horns of a dilemma with regard to their accountabilities as they juggle competing demands and responsibilities. More than a decade ago, Charles J. Austin reminded health administrators that they must equitably balance multiple accountabilities in successfully managing health care institutions ([3], p. 23). In doing so, they must respect the reality that multiple interest groups will attempt to influence their decisions, such as private investors, the community, patients, various governmental and regulatory bodies, multiple business or labor organizations, or third party payers. Being charged with the administration of the institution, and in large part, the determination of the health care that will ultimately be delivered, administrators must be able to mediate a system filled with conflicts and opposing demands. Nonetheless, we argue that the fundamental institutional mission of beneficence and social justice should be their guides.

Toward this end, three general recommendations are offered:

1. The recognition of the traditional healing mission and the charitable ethic must not be lost. While compensatory efforts to ensure greater efficiency and better control are necessary, institutions and their administrators must never lose sight that the principal purpose of health care providers is to heal those in need within the limit of the resources available.

2. The consideration of the ethical implications of administrative decisions must not be denied, ignored or minimized. Management often leans toward strictly financial solutions, acknowledging that these may be the most straightforward and may pose what appears to be the best short-term solution for the institution. Such short-range financial strategies often carry serious social and ethical implications, particularly for the most needy in our society. Administrators should gain a fuller appreciation of what constitutes an ethical problem and how common administrative actions may precipitate ethical dilemmas. Further, their knowledge of certain fundamental ethical principles such as beneficence, justice, and respect for persons could facilitate their making sound decisions [18].

3. The development of a system to ensure charity and access for the least well-off members of a society is essential and must be compassionately supported by all members of the health care industry. It is not impossible to maintain a charitable ethic and succeed as a business enterprise; however, such a goal will not be achieved unless a collective effort is undertaken by all sectors of the health care industry.

In sum, it is clear that the ethical principles and values in health administration shall face increasing challenge in the years ahead. A growing business orientation and an extremely competitive environment threatens the historic humanistic ethos. The advent of health care for profit possibly portends a lessening or even an abandonment of cherished health care traditions. Institutional survival and a strong "bottom line" are the contemporary driving forces overcoming the traditional emphasis of caring and compassion for the infirm, giving charity to the poor and less fortunate, and ensuring quality of care for all at any cost.

The extent to which such a fundamental shift will stimulate administrators to reassess their values remains unknown. Yet, in the midst of such change, continued excellence in American health care demands an ongoing examination of the ethical implications of hospital decisions. While the prevailing trend has gained widespread acceptance as a necessary response to current economic and financial realities, more fundamentally, one must question whether it more accurately reflects a dramatic change in the values of our society.

University of New Hampshire
Durham, New Hampshire

NOTE

* Work on this essay was supported by a grant from the PEW Memorial Trust through the Commission on Ethical Issues in Health Management of the Association of University Programs in Health Administration.

BIBLIOGRAPHY

[1] American College of Hospital Administrators: 1980, *Code of Ethics*, American College of Hospital Administrators, Chicago.

[2] American College of Hospital Administrators and American Hospital Association: 1941, 1957, *Code of Ethics*. American Hospital Association, Chicago.

[3] Austin, C. J.: 1974, 'What is Health Administration?', *Hospital Administration 19*, 14–29.

[4] Beauchamp, T. L. and McCullough, L. B.: 1984, *Medical Ethics: The Moral Responsibilities of Physicians*, Prentice-Hall, Inc., Englewood Cliffs, New Jersey.

[5] Bloom, B.: 1981, 'Utilization Patterns and Financial Characteristics of Nursing Homes in the United States 1977', National Center for Health Statistics, Hyattsville, Maryland.

[6] Corwin, E. H.: 1946, *The American Hospital*, Commonwealth Fund, New York.

[7] Cunningham, R. M. Jr.: 1982, *The Healing Mission and the Business Ethic*, Pluribus Press, Chicago.

[8] Cunningham, R. M. Jr.: 1983, 'More Than a Business: Are Hospitals Forgetting Their Basic Mission?', *Hospitals 57*, 88–90.

[9] DeGeorge, R. T.: 1982, 'The Moral Responsibility of the Hospital', *Journal of Medicine and Philosophy 7*, 87–100.

[10] Dowling, W. L.: 1984, 'The Hospital', in S. J. Williams and P. R. Torrens (eds.),

Introduction to Health Services, 2nd edition, John Wiley & Sons, New York, pp. 172—215.

[11] Flexner, A.: 1910, *Medical Education in the United States and Canada* (The Flexner Report), The Carnegie Foundation for the Advancement on Teaching, reprinted: Science and Health Publications, Bethesda, Maryland.

[12] Freymann, J. G.: 1974, *The American Health Care System: Its Genesis and Trajectory*, Medcom Press, New York.

[13] Freedman, S.: 1985, 'Megacorporate Health Care', *New England Journal of Medicine 312*, 579—582.

[14] Friedman, E.: 1982, 'The Dumping Dilemma: The Poor Are Always With Some of Us', *Hospitals 56*, 51—56.

[15] Friedman, E.: 1982, 'The 'Dumping' Dilemma: Finding What's Fair', *Hospitals 56*, 75—84.

[16] Gray, B. H.: 1983, 'An Introduction to the New Health Care for Profit', in B. H. Gray (ed.), *The New Health Care for Profit: Doctors and Hospitals in a Competitive Environment*, Institute of Medicine, National Academy Press, Washington, D.C.

[17] Hiller, M. D.: 1981, 'Medical Ethics and Public Policy', in Marc D. Hiller (ed.), *Medical Ethics and the Law: Implications for Public Policy*, Ballinger Publishing Company, Cambridge, Massachusetts.

[18] Hiller, M. D.: 1984, 'Ethics and Health Care Administration: Issues in Education and Practice', *Journal of Health Administration Education 2*, 148—190.

[19] Kinzer, D. M.: 1984, 'Care of the Poor Revisited', *Inquiry 21*, 5—16.

[20] Knox, R. A.: 1984, 'Some Local Hospitals "Dump" the Uninsured', *Boston Globe*, February 6, p. 31.

[21] Johnson, D. E. L.: 1985, 'HCA, American Agree to Form $8.5 Billion Giant', *Modern Healthcare 15*, 16—22.

[22] Johnson, D. E. L.: 1985, 'Investor-owned Chains Continue Expansion, 1985 Survey Shows', *Modern Healthcare 15*, 75—110.

[23] Lewis, L. S., *et al.*: 1981, 'Investor-owneds and Nonprofits Differ in Economic Performance', *Hospitals 55*, 52—58.

[24] MacEachern, M. T.: 1957, *Hospital Organization and Management*, Third Edition, Physician's Record Company, Chicago.

[25] Neuhauser, D.: 1983, *Coming of Age: A 50-Year History of the American College of Hospital Administrators and the Profession It Serves — 1933-1983*, Pluribus Press, Chicago, pp. 73—84.

[26] Norman, D. K.: 1985, 'Managing Interhospital Patient Transfers', *Hospitals 55*, pp. 98—99, 101.

[27] Pattison, R. V. and Katz, H. M.: 1983, 'Investor-Owned and Not-for-Profit Hospitals: A Comparison Based on California Data', *New England Journal of Medicine 309*, 347—353.

[28] O'Rourke, K.: 1984, 'An Ethical Perspective on Investor-Owned Medical Care Corporations', *Frontiers of Health Services Management 1*, 10—26.

[29] Pellegrino, E. D.: 1978, 'Medical Morality and Medical Economics: The Conflict of Canons', *Hastings Center Report 8*, 8—12.

[30] Pellegrino, E. D. and Thomasma, D. C.: 1981, *A Philosophical Basis of Medical Practice: Toward a Philosophy and Ethic of the Health Professions*, Oxford University Press, New York and Oxford.

[31] Pellegrino, E. D.: 1985, 'Catholic Hospitals: Survival Without Moral Compromise', *Health Progress 66*, 42—49.

[32] Relman, A. S.: 1980, 'The New Medical-Industrial Complex', *New England Journal of Medicine 303*, 963—970.

[33] Relman, A. S.: 1983, 'Investor-Owned Hospitals and Health Care Costs', *New England Journal of Medicine 309*, 370—372.

[34] Rosen, G.: 1963, 'The Hospital: Historical Sociology of a Community Institution', in E. Friedson (ed.)., *The Hospital in Modern Society*, The Free Press, New York, pp. 1—36.

[35] Rosett, R. N.: 1984, 'Doing Well by Doing Good: Investor-Owned Hospitals', *Frontiers of Health Services Management 1*, 2—9.

[36] Shore, M. F. and Levinson, H.: 1985, 'On Business and Medicine', *New England Journal of Medicine 313*, 319—321.

[37] Starr, P.: 1982, *The Social Transformation of American Medicine*, Basic Books, Inc., New York.

[38] State of Florida: 1983, *Hospital Cost Containment Board, Annual Reports, 1981-82, 1982-83.* Florida Hospital Cost Containment Board, Tallahassee, Florida.

[39] Stern, B. J.: 1946, *Medical Services* by *Government: Local, State, and Federal*, The Commonwealth Fund, New York.

[40] Thomasma, D. C.: 1982, 'Hospitals' Ethical Responsibilities as Technology Regulation Grow', *Hospital Progress 63*, 74—79.

[41] Young, Q. D.: 1984, 'Impact of For-Profit Enterprise on Health Care', *Journal of Public Health Policy 5*, 449—452.

[42] Young, Q. D.: 1985, 'The Danger of Making Serious Problems Worse', *Business and Health 2*, 32—33.

LU ANN ADAY

SHIFTING THE PRIORITIES AND VALUES: A COMMENTARY ON HILLER AND GORSKY

In the ancient folk tale of 'The Blind Men and the Elephant', three blind men were asked to touch an elephant and then describe what it resembled. One man touched its side and said it was like a massive wall. Another, putting his arms around its leg, said it resembled a tall and sturdy tree. The third, grasping its trunk, argued that it must surely be slender and serpentine like a snake. Similarly, our own perspectives on problems often are shaped by the particular aspects we choose to see and by those which fall outside our particular views of the problems.

Though in their paper, 'Shifting Priorities and Values: A Challenge to the Hospital's Mission', Marc Hiller and Robin Gorsky have focused in a provocative way on the traditional and emerging emphases of hospitals and hospital administrators on the business versus service orientation, their discussion tends to belie other perspectives on the same set of historical facts about the evolution of hospitals and to omit other major issues, actors, or evidence in considering the priorities and values implicit in defining the modern hospital's mission. In this paper I discuss some of these alternative foci and suggest that the authors set up a false dichotomy regarding the alternative motivations that may drive the operation of the health care system in the United States in the future: "to maximize the quality of care and availability of resources for the population *or* maximize the return on investment . . . ."

The way this question is posed implies a framework for examining the business and service orientations of hospitals as either-or polarities, rather than as a matrix of perhaps competing, but simultaneous, institutional objectives. It is perhaps more accurate to argue, both historically and across types of institutions at the present time (for example, with regard to profit and non-profit hospitals), that each of these objectives has been or is an aspect of the hospital mission to a greater or lesser degree and that a major goal of many institutions and providers is how to maximize the realization of the access, quality, *and* cost-containment objectives simultaneously.

Odin W. Anderson and Norman Gevitz, in an historical overview of the development of the modern hospital, argue that the "hospital . . .

G. J. Agich and C. E. Begley (eds.), The Price of Health, 263–272.

has been to a considerable extent, a mirror of the world around it" ([2], p. 316). The mirror that those authors offer to explain the evolution of this institution, however, appears to be more multi-faceted than Hiller and Gorsky provide. Hiller and Gorsky seem to convey that hospitals have always been devoted to the "charitable healing mission". Anderson and Gevitz, on the other hand, trace the development of the institution from its earliest historical precursors in the Greek temples to the cult of Aesculapius, to the Roman *valetudinaria,* which cared for soldiers and others whose recovery was considered to be in the best interests of the state, to much later, with the rise of Christianity, the shift from serving the interests of the empire to serving the values of charity taught by the new religion. Although the earliest modern-day hospitals did come to emphasize "charity" as a principal goal, it is not clear whether they were *de facto* as successful in realizing the "healing" and "patient beneficence" missions as Hiller and Gorsky imply. In fact, Anderson and Gevitz point out the less-than-idealistic nineteenth-century attitude toward the hospital:

> Throughout the greater part of the nineteenth century, the general hospital did not enjoy an enviable reputation among the public nor did it deserve one. The upper and middle classes viewed it as an institution exclusively for the poor, while the lower strata, which it served, regarded hospitalization as stigmatizing — a sign of family breakdown and economic failure. More important, most people, rich and poor, considered hospitals to be death houses. Indeed, mortality was high even for seemingly minor ailments — this the result of the constant threat of infection passed from patient to patient with the physician often serving as a transmitter ([2], p. 309).

In his book, *The Social Transformation of American Medicine,* Paul Starr further claims that the emergence of the interest in hospitals as businesses is not as recent a phenomenon as Hiller and Gorsky seem to imply. He points out that "hospital care turned into a considerable industry at the end of the nineteenth century . . . . And large numbers of new hospitals were established, the majority as business enterprises" ([24], p. 148). Around the turn of the century, in fact, more than half of the hospitals in the United States were for-profit proprietary institutions, established by one or a small group of physicians to have a place to hospitalize their patients. The major trend in recent years has not been the growth of for-profit hospitals *per se,* but rather the purchase of hospitals by investor-owned corporations to form for-profit multi-hospital *systems* ([5], pp. 182—183; [24]). A more refined focus on the part of the authors in examining the evolution and implications of the for-profit side of the modern hospital industry, then, would have been on the emergence of these corporate entities in particular.

In introducing the various levels that have been or should be considered in addressing the major issues confronting the health care system in the United States, a significant omission on the part of the authors, it seems, between the messo (institutional) and macro (societal) level is the *community* served by the modern hospital. A hospital's mission and the foci that are influential in shaping it are determined to a considerable extent by whom it chooses to (or must) serve and the immediate political and social environment in which it operates [21]. "Community" hospitals (non-federal short-term hospitals, regardless of whether they are for-profit or non-profit or single entities or part of a multi-hospital system) represent more than 80 percent of the nation's hospitals and provide care to more than 90 percent of all patients admitted to hospitals. There is evidence of increasing pressure for these institutions to become even more responsive to the multiplicity of health care needs in the communities they serve — such as in the provision of prenatal care clinics for pregnant teenagers, home health services for the elderly, or twenty-four-hour urgent care centers for young middle-class families ([5], pp. 184—185). The authors seem to disregard the importance of these factors in potentially mediating the ability of individual administrators to achieve their goals or to consider the significant role the community may play in both defining and enabling the implementation of a given institution's expressed mission [13].

In addition to omitting consideration of the community served by the hospital in considering the forces impacting on the modern hospital and the administrators who attempt to manage it, the authors fail to consider adequately the role of the institution's governing board in formulating and implementing its goals as well. The governing board is, in a sense, an institutional representation of the community served by the hospital. In a recent article entitled, "Ethics and Health Administration", D. K. Oglesby, Jr., suggests that trustees often have been one of the principal proponents of the community service orientation [15]. Further, the increased *institutional* responsibility being thrust upon hospitals, such as that enforced by the *Darling v. Charleston* legal case establishing the concept of corporate or institutional responsibility for ensuring the quality of patient care, increasingly is being assumed by hospitals' governing boards, rather than by hospital administrators per se ([5], p. 191). Unfortunately, Hiller and Gorsky do not address the role that trustees might play either in mediating the dominance of the business orientation or in enforcing patient-oriented standards of practice in many community-based institutions.

Turning to the internal structure of hospitals, the authors assert that "little argument exists that both hospitals and health administrators traditionally have espoused a commitment to a 'healing' mission similar to that of physicians and other providers". They do not, however, provide any direct evidence for this assertion. On the contrary, there is considerable evidence that the emphasis more often has been on the dual authority structure that exists in hospitals, with administrators more often having been responsible for the efficient management of the institution as a whole and the medical staff more directly concerned with the "healing" or patient care functions. It is this dynamic tension, rather than the necessary complementarity, between the physicians' and administrators' roles which has probably more often characterized their interactions in resolving moral conflicts [12].

Hiller and Gorsky also seem to portray an ideal of the hospital administrator as a kind of superior "moral agent" — an institutional Lone Ranger who is supposed to uphold the rights of the patients and enforce the mission of the institution. Perhaps the most effective emphasis might be, as Oglesby and others have argued, to strive "to balance a service commitment with a hard-headed business approach" ([15], p. 39) to running a hospital and that the most effective mechanism to accomplish this objective would be to strive for a full partnership and consensus among the administrator, the board, medical staff, hospital employees and the community served. James W. Summers [25] and Laurence B. McCullough [12] further point out the particular contributions that a multidisciplinary ethics committee could make to mediating difficult moral choices within the institution. Interestingly, Hiller and Gorsky do not discuss the role of this increasingly important mechanism for resolving potential conflicts of values. An alternative to relying principally on the moral leadership of the administrator, which Hiller and Gorsky seem to stress, then, is continuing to strive for an effective working *partnership* among the variety of actors concerned with defining and carrying out the modern hospital's mission [23].

The authors devote a considerable effort to discussing the various ethical codes that have been developed to govern the behavior of hospitals and hospital administrators. On the other hand, Oglesby argues that "these codes of ethics are useful only for the honest and well-meaning individual, for those who make an effort to do the right thing and are guided by such a code. For those who do not care, the code of ethics is meaningless" ([15], p. 41). In order to encourage those administrators who do care to act ethically, a starting point might be to consider working with physicians, trustees and ethics committees to formulate a set of institutional values and objectives that could serve as norms or guidelines for shaping and/or enforcing agreed-upon precepts within the institution at least [13].

In their discussion of the 'Ethical Implications of the Growth in the For-Profit Sector', the authors provide a litany of abuses they expect to result from the fact of the growth of this sector. However, they offer little or no empirical evidence to support their assertions. Many statements seem to reflect an assumption of some "golden age" of health care that now will be jeopardized by these developments. There is no effort to hypothesize where there might be positive, as well as negative, outcomes resulting from these changes.

In fact, the literature suggests a variety of both positive and negative consequences regarding access, cost containment, and quality objectives. They are offered here as a more "balanced" perspective than that provided by Hiller and Gorsky, in hypothesizing about the health care system to emerge from these changes.

Access

On the negative side, in terms of access, it has been suggested, as Hiller and Gorsky also imply, that because of the profit-oriented motives of corporate practice and/or the cost ceilings imposed by prospective payments, institutions may be less willing to care for certain types of patients. A recent article on one major for-profit hospital system (Humana), for example, explained:

> Private insured patients can be charged what the market will bear. When a hospital has empty beds, Medicare and Medicaid patients are better than cold sheets, and Humana treats as few of those patients as possible. Humana prefers to own facilities in suburbs where young working families are having lots of babies. Though young people use hospitals less than the elderly, they are more likely to be privately insured and in need of surgery, which makes the most money. The babies provide a second generation of customers ([24], p. 436).

Or, in a guide to ethical decision making, an example relating to the impact of external factors on such decisions argued:

> An inadequate level of Medicare reimbursement never justifies inadequate care to patients; it may, however, be an important consideration in deciding whether to limit the number of such patients one can serve ([10], p. 177).

Another adaptation on the part of cost-conscious corporate arrangements might be to develop a two- or three-tiered system of care in which uninsured or publicly insured individuals receive less intensive or comprehensive care than fully insured or high income patients ([24], p. 448). There is evidence that patients are less satisfied in highly bureaucratized settings or those in which there is not a clearly defined linkage with one particular physician [6], [7]. This may be exacerbated with the advent of more formalized doctor-patient arrangements. A concern with the advent of diagnosis-related standardization of reimbursements is that there is less discretion about delaying the discharge of patients for which there are inadequate post-hospitalization care arrangements [20].

On the other hand, there are arguments as well that access might be enhanced by newly emerging organizational and financial changes. As providers increase their competition for patients, they may improve the convenience or range of services offered to capture a larger share of the market. Marketing itself, it might be argued, serves to improve the information available to potential consumers about which provider might best meet their needs. A product of the diversification of the health care market is the emergence of new forms of care, for example, hospices, which may, in fact, be more "humane" and satisfactory caretaking models to the patient and his family. The division of labor increasingly emerging in medicine, in which para-medical personnel perform procedures traditionally performed by physicians — for example, general exams, pap smears, screening tests, could mean that patients have quicker access to "experts", than if they had to wait to see the physician himself [26].

Cost Containment

In terms of the cost containment objective, questions have been raised about whether corporate-based medicine or prospective payment are really the best cost-containing alternatives. That is, do they really work? For example, there is some evidence that joining a multihospital chain may mean that, at least initially, the costs of a hospital's operation increase or that there will be significant overhead costs associated with putting prospective reimbursement systems in place such as changes in recordkeeping, addition of personnel, etc., which will minimize the actual cost savings. Others argue that there will be a tendency toward "cream-skimming", that is, providing more of those types of services which yield the biggest profit margin [16]. Critics of competition assert that physicians will find other ways to get around reduced income resulting from reimbursement constraints through increasing fees or ordering more of certain types of tests and procedures. Some say that cost saving in health care, in fact, can be realized more directly by investing in other sectors such as housing, occupational safety, nutrition, or the environment where the ultimate payoffs to "health" may be even greater.

Those who think these newly emerging forms will work better in achieving the cost containment goal assert that corporate modes of practice can capitalize on economies of scale and other efficiencies resulting from a more horizontally — or vertically — integrated form of organization [22]. Product diversification may also mean that there can be more programs dealing directly with health maintenance or prevention that may, in fact, yield larger returns in the long run than dealing almost exclusively with illness-related care. The professionalization of administrators charged with running more complex health care organizations may mean they are better "managers" of institutional resources. Further, physician behavior is the "black box" where most of the decisions relating to the use of resources are made. To the extent that physicians become more involved in the matrix of decision making dealing with both cost *and* quality concerns, the greater potential there is for actually effecting informed cost-saving decisions.

Quality

A major criticism that is offered of the forms of corporate organization and financing is that the quality of care may be seriously compromised. Implied in the criticism of these developments is diminished autonomy on the part of physicians in making patient care decisions. More often, third parties — institutional or corporate review boards, other physicians, or insurance auditors — could intervene to set constraints or parameters on the individual providers' decision making. The Hippocratic injunction to "do no harm" may be compromised if cost, rather than quality, becomes a dominant criterion in monitoring patient care. There may be a tendency to discharge patients too soon or to hold back in administering certain tests or procedures to save costs. These results may "backfire" both to the individual provider and the institution in terms of increased malpractice claims or an overall diminished quality of patient care ([20], [24]).

Some, however, would argue that DRGs better enforce the norm of similar treatment for similar cases, in that variant forms of practice — which may differ in cost as well as efficacy — become more "bounded" by more standardized modes of practice. These new constraints may, in fact, reduce the number of unnecessary tests and procedures. More is not necessarily better because modern health care creates as well as cures some problems through over-prescribing, exposing patients to the risks of other illnesses or accidents through hospitalization, or failing to coordinate multifaceted care programs ([3], [9]).

These, then, are some of the ethical questions relating to the system goals of access, cost, and quality that arise in the context of newly emerging forms of organizing and financing care. Many of these are in fact empirically testable "hypotheses". They should be viewed as just that, however — hypotheses, which are subject to being confirmed or disconfirmed by the weight of evidence not yet readily available.

Dramatic changes are underway in the health care system as a whole and in the hospital industry in particular. A careful look at history, however, seems to reveal that these issues are recurring, rather than new ones. Accordingly, a careful look at the society and community in which hospitals operate may help us to understand the deeper origins of the "crises" confronting them at the present time. Acknowledging the role that a *variety* of actors plays in shaping the institution may lead to a more constructive effort to design solutions to difficult problems. This requires, however, being willing to look at the "what if's" in terms of potentially positive *and* negative outcomes. Such an approach will enable a fairer evaluation of the emerging alternatives and consideration of the *balance* of objectives which individual hospitals (or the United States health care system as a whole) now are struggling to achieve.

School of Public Health
University of Texas Health Science Center
Houston, Texas

BIBLIOGRAPHY

[1] Altman, D.: 1983, 'Health Care for the Poor', *Annals of the American Academy of Political and Social Sciences 468,* 103—121.

[2] Anderson, O. W. and Gevitz, N.: 1983, 'The General Hospital: A Social and Historical Perspective', in D. Mechanic (ed.), *Handbook of Health, Health Care, and the Health Professions,* The Free Press, New York, pp. 305—317.

[3] Carlson, R. J.: 1975, *The End of Medicine,* John Wiley & Sons, New York.

[4] Davidson, S.: 1982, 'Physician Participation in Medicaid: Background and Issues', *Journal of Health Politics, Policy and Law 6,* 703—717.

[5] Dowling, W. L.: 1984, 'The Hospital', in S. J. Williams and P. R. Torrens (eds.), *Introduction to Health Services,* John Wiley & Sons, New York, pp. 172—215.

[6] Fleming, G. V.: 1981, 'Hospital Structure and Consumer Satisfaction', *Health Services Research 16,* 43—63.

[7] Freidson, E.: 1970, *Profession of Medicine*, Harper & Row, Publishers, New York.

[8] Goldsmith, J. C.: 1981, *Can Hospitals Survive? The New Competitive Health Care Market,* Dow Jones-Irwin, Homewood, Illinois.

[9] Illich, I.: 1976, *Medical Nemesis: The Expropriation of Health,* Bantam Books, Toronto.

[10] Jonsen, A. R. *et al.*: 1982, *Clinical Ethics,* Macmillan Publishing Co., Inc., New York.

[11] Komaroff, A. L.: 1983, 'The Doctor, the Hospital, and the Definition of Proper Medical Practice', in President's Commission for the Study of Ethical Problems in Medicine and Biomedical and Behavioral Research, *Securing Access to Health Care: The Ethical Implications of Differences in the Availability of Health Services. Vol. Three: Appendices,* pp. 225—251.

[12] McCullough, L. B.: 1985, 'Moral Dilemmas and Economic Realities', *Hospital and Health Services Administration 30,* 63—75.

[13] McNerney, W. J.: 1985, 'Managing Ethical Dilemmas', *Journal of Health Administration Education 3,* 331—340.

[14] Mitchell, J. B. and Cromwell, J.: 1980, 'Medicaid Mills: Fact or Fiction', *Health Care Financing Review 2,* 37—49.

[15] Oglesby, D. K., Jr.: 1985, 'Ethics and Hospital Administration', *Hospital & Health Services Administration 30,* 29—43.

[16] Pattison, R. V. and Katz, H. M.: 1983, 'Investor-owned and Not-for-profit Hospitals', *New England Journal of Medicine 309,* 347—353.

[17] Rogers, S. J., *et al.*: 1985, *Hospitals and the Uninsured Poor: Measuring and Paying for Uncompensated Care,* United Hospital Fund of New York, New York.

[18] Salmon, J. W.: 1985, 'Profit and Health Care: Trends in Corporatization and Proprietization', *International Journal of Health Services 15,* 395—418.

[19] Sauer, J. E., Jr.: 1985, 'Ethical Problems Facing the Healthcare Industry', *Hospitals and Health Services Administration 30,* 44—53.

[20] Shakno, R. J. (ed.): 1984, *Physician's Guide to DRGs,* Pluribus Press, Chicago.

[21] Shortell, S. and Kaluzny, A.: 1983, *Health Care Management: A Text in Organization Theory and Behavior,* John Wiley & Sons, New York.

[22] Sloan, F. A. and Vraciu, R. A.: 1983, 'Investor-owned and Not-for-profit Hospitals: Addressing Some Issues', *Health Affairs 2,* 25—37.

[23] Spivey, B. E.: 1984, 'The Relation Between Hospital Management and Medical Staff Under a Prospective Payment System', *New England Journal of Medicine 310,* 984—986.

[24] Starr, P.: 1982, *The Social Transformation of American Medicine,* Basic Books, Inc., New York.

[25] Summers, J. W.: 1985, 'Closing Unprofitable Services: Ethical Issues and Management Responses', *Hospital & Health Services Administration 30,* 8—28.

[26] Tarlov, A. R.: 1983, 'Shattuck Lecture: The Increasing Supply of Physicians, the Changing Structure of the Health Services System, and The Future Practice of Medicine', *New England Journal of Medicine 308,* 1235—1244.

NOTES ON CONTRIBUTORS

Lu Ann Aday, Ph.D., is Associate Professor of Behavioral Science in the School of Public Health, University of Texas Health Science Center, Houston, Texas.

George J. Agich, Ph.D., is Associate Professor in the Departments of Medical Humanities, Philosophy, and Psychiatry and Director of the Ethics and Philosophy of Medicine Program, Southern Illinois University School of Medicine, Springfield, Illinois.

Robert Audi, Ph.D., is Professor in the Department of Philosophy at the University of Nebraska, Lincoln, Nebraska.

Mary Ann Baily, Ph.D., is Adjunct Associate Professor in the Department of Economics at George Washington University, Washington, D.C.

Charles E. Begley, Ph.D., is Assistant Professor of Economics in the School of Public Health, University of Texas Health Science Center, Houston, Texas.

David D. Friedman, Ph.D., is Associate Professor of Economics in the A.B. Freeman School of Business, Tulane University, New Orleans, Louisiana.

Robin D. Gorsky, Ph.D., is Assistant Professor in the Department of Health Administration and Planning, School of Health Studies, University of New Hampshire, Durham, New Hampshire.

Marc D. Hiller, Dr.P.H., is Associate Professor in the Department of Health Administration and Planning, School of Health Studies, University of New Hampshire, Durham, New Hampshire.

Albert R. Jonsen, Ph.D., is Professor of Ethics in Medicine in the Department of Medicine, School of Medicine, University of California, San Francisco, California.

Michael M. Kaback, M.D., is Professor of Pediatrics and Medicine and Associate Chief, Division of Medical Genetics, U.C.L.A. School of Medicine, Harbor-U.C.L.A. Medical Center, Los Angeles, California.

Paul T. Menzel, Ph.D., is Professor of Philosophy at Pacific Lutheran University, Tacoma, Washington.

E. Haavi Morreim, Ph.D., is Assistant Professor in the Department of Human Values and Ethics, College of Medicine, University of Tennessee Health Sciences Center, Memphis, Tennessee.

Marc J. Roberts, Ph.D., is Professor of Political Economy and Health Policy in the Department of Health Policy and Management, School of Public Health, Harvard University, Cambridge, Massachusetts.

Stuart F. Spicker, Ph.D., is Professor of Community Medicine and Health Care, School of Medicine, University of Connecticut Health Center, Farmington, Connecticut.

J. Michael Swint, Ph.D., is Professor of Economics, School of Public Health, University of Texas Health Science Center, Houston, Texas.

Peter S. Wenz, Ph.D., is Associate Professor of Philosophy at Sangamon State University, Springfield, Illinois.

Gerald R. Winslow, Ph.D., is Professor of Religion at Walla Walla College, College Place, Washington.

INDEX